ENVIRONMENTAL SUSTAINABILITY IN THE 21ST CENTURY

EMERGING ISSUES AND THE WAY FORWARD

DR. RANJAN SARKAR

Environmental Sustainability in the 21st Century

Emerging Issues and the Way Forward

Environmental Sustainability in the 21st Century

Emerging Issues and the Way Forward

Editor

Dr. Ranjan Sarkar

M.A. (Geography & Applied Geography)
Gold Medalist, B.Ed., M.Ed., N.E.T., Ph.D.
Assistant Professor, Ananda Chandra Training College
(Govt. of West Bengal Sponsored Teachers' Training College)
Jalpaiguri, West Bengal, India
Email: ranjangeo@gmail.com
Website: www.drranjansarkar.in

NAMYA PRESS.com

Published by
Namya Press
213, Vardhan house, 7/28, Ansari Road,
Darya Ganj, Delhi - 110002.
Email: namyapress@gmail.com

Edition	:	First Published in 2020 (Namya Press)
Title	:	Environmental Sustainability in the 21st Century Emerging Issues and the Way Forward
Editor	:	Dr. Ranjan Sarkar
ISBN	:	978-81-947715-1-7
MRP	:	₹ 1495.00
Copyright	:	© Dr. Ranjan Sarkar 2020 All rights reserved
Printer	:	Trident Enterprises (Noida), D-204, Sector-10, Noida G.B. Nagar, U.P.

This Book is Dedicated

to

The Covid-19 Warriors of India

Professor D. K. Mandal, Ph.D.
Professor of Physical Geography,
Department of Geography & Applied Geography,
University of North Bengal (Accredited by NAAC with Grade A)
Siliguri, West Bengal, India.
Email: dkmandalnbu@gmail.com

Foreword

Global environmental change is as old as planet Earth. Physical, chemical, and biological processes have been shaping and reshaping the earth's environment since its infancy 4.5 billion years ago. In recent time, however, humankind has been one of the major driving forces of change on the planet, including climate change, stratospheric ozone depletion, and deforestation, loss of biodiversity, acidification, pollution, and desertification.

Nature and natural environments can be immensely varied, and range from a neatly structured and well-planned garden to the great outdoors of untouched mountains and large forest areas. The majority of the global population lives in cities, and the rate of urbanization is increasing; urban nature thus becomes an important topic in terms of population health. Research into how environmental change shapes behaviour patterns provides an important background for interventions designed to change habits regulated by impulsive, rather than reflective, regulatory processes. We perceive aspects of the environment that, in the past, have been associated with rewarding experiences and we respond to repeat those experiences. Over time this process of operant conditioning establishes automatic responses to particular environmental stimuli, or cues, which, through sequencing, can be built up into complex impulsively regulated behaviour patterns. Such analyses models can be applied to identify potentially effective change techniques capable of modifying impulsive processes (in stage 3 of IM).

In the light of the above, it is a welcome initiative to bring a publication on "Environmental Sustainability in the 21st Century: Emerging Issues and the Way Forward" with 24 research articles edited by Dr. Ranjan Sarkar. The book covers with 3 sub-themes, namely, 'Natural Environment and Sustainability' with 6 research papers, 'Environmental Challenges in the Present Context' with 10 research papers, and 'Environmental Management and Practices' with 8 research papers.

The selected and edited articles are very interesting in that each of them typically concerns different issues and concerns in the fields of environmental sustainability. The research article on excessive rainfall and resultant urban flooding is an important study in the field of applied geomorphology. The fluctuations of groundwater level in Rajasthan based on spatio-temporal data is an outstanding contribution. The insect biodiversity in one more very interesting presentation. The present scenario in Indian Sundarbans has been presented taking due consideration of biodiversity. The research article on super cyclone Amphan concerns with impacts on environment and particularly on global warming and climate change. The effect of microclimate on Tea cultivation is an outstanding contribution. The issues related to Covid-19 Crisis has been explained with utmost care. The research article on the problem of immigration in Assam and its impact on economy is an important issue for long. How the environmental degradation affecting livelihood has been explained nicely. The problem of river pollution in India is an interesting study in environmental geography. The research article on river bank erosion induced human displacement is an outstanding contribution in the field of settlement geography. The arsenic contamination of groundwater in West Bengal is a nice contribution in the field of applied geomorphology. The study on vegetation and water estimation by using modern geospatial tools is one more nice contribution. An attempt has been made to establish a linkage between Chemistry and Geography. The research article on estimation of groundwater resources is an outstanding contribution in the field of environmental geography. Water crisis in off-monsoon period in India has already drawn our attention. The malnutrition is one more global issue particularly in desert and hilly terrain. The research article on an assessment of the status of household solid waste and its management is one of the quality works in the field of applied geography.

I do firmly believe that the research articles will give new directions to research in applied geography and inspirations to the practicing geographers, viz., and the community of young serious researchers. I take this opportunity to thank Dr. Ranjan Sarkar and wish a very high prospect for this publication.

D K Mandal

Professor D. K. Mandal., Ph.D.

Editor
Dr. Ranjan Sarkar
Assistant Professor

Preface

The existence of life on the earth is in danger. Environmental degradation has threatened the survival of human beings and other species. Human beings have much harmed to the natural environment through their actions. Increasing population, urbanization and poverty have further created pressure on the natural resources and led to the degradation of the environment. The concern for the environment is getting diminished among the people. The world is going through an environmental crisis which has drawn the attention of environmentalists, intellectuals and policy makers throughout the world. Environmental degradation, pollution, ozone depletion, deforestation, climate change, loss of biodiversity, global warming, ecological imbalance, unsustainable development etc. are some of the frightening challenges that may threaten our future, if we do not face up to them. These challenges need to be addressed seriously and sincerely. It should be remembered that the Earth does not belong to us, we belong to the Earth.

Human societies are nothing, but subsystems of the biosphere. Protecting and preserving the environment is essential for the conservation of nature, biosphere and the existence of life on the Earth. It is one of the basic prerequisites for the overall development of any country in the world. It is the need of the hour to protect and preserve the environment for the present as well as future generations to come. There is a requirement to save resources and reconnect with the nature to save the planet. Sustainable development should be facilitated to conserve and preserve the resources. The idea of environmental sustainability can be used as a tool to conserve the ecosystem. It is concerned with the interaction with the environment to avoid depletion or degradation of natural resources and allow for long-term environmental quality. It focuses on renewable resources and protecting the long-term productivity to meet the future social and economic needs.

The food we eat, the air we breathe, the water we drink and the climate that makes our planet habitable all come from nature. Nature is the driving and nurturing force of life on the Earth. It is an integral part of our environment. It plays a significant role in our existence. Natural environment influences all the aspects of life directly or indirectly. Protection and preservation of nature is the urgent need of the 21st century. It is our duty to respect, safeguard and conserve the nature to meet the needs of the present as well as future generations. The idea of sustainable development can play a crucial and critical role in the conservation of nature. Masses should be made aware of reducing wastes, reusing resources and recycling materials. It is our duty to take care of our environment and ecosystem. In order to meet the challenges of environment and development, there is a need of genuine cooperation, solidarity and partnership amongst the people. Earth is not just a planet. It is our home.

In this book "Environmental Sustainability in the 21st Century: Emerging Issues and the Way Forward" total 24 research articles are there and the papers have been grouped into three sections, each of which is highly relevant in the present-day context. The first part of this volume entitled 'Natural Environment and Sustainability' is concerned with the issues related with the natural environment and its sustainability with 6 research papers. The second part is related with 'Environmental Challenges in the Present Context' with 10 research papers, and the third part 'Environmental Management and Practices' is encompassing with 8 research papers mostly dealing with sustainable environmental management practices. These are very significant aspects of environmental sustainability in this 21st Century. This is an honest effort by me to explore the new dimensions of Environmental Sustainability in this 21st Century and hope this will be of great use for the researchers, academicians, policymakers and stakeholders.

Place: Siliguri

Date: 16/07/2020

Dr. Ranjan Sarkar, Editor

M.A. (Geography & Applied Geography)
Gold Medalist, B.Ed., M.Ed., N.E.T.,
Ph.D.(Geography & Applied Geography),
Assistant Professor in Geography,
Ananda Chandra Training College,
Jalpaiguri, West Bengal, India.
Email: ranjangeo@gmail.com
Website: www.drranjansarkar.in

Acknowledgements

"Environmental Sustainability in the 21st Century: Emerging Issues and the Way Forward" is an attempt by me to highlight some of the major issues of 21st Century Environmental Sustainability. First of all, I would like to thank my respected teacher Professor Dipak Kumar Mandal, Professor of Physical Geography, Department of Geography & Applied Geography, University of North Bengal (Accredited by NAAC with Grade A), Siliguri, West Bengal, India, for his all positive attitude and his support by contributing the Foreword of this book.

I am indebted to my respected teacher Prof. (Dr.) Ranjan Roy, Department of Geography and Applied Geography, University of North Bengal for his visionary support throughout the process. I am grateful to my respected teachers Prof. Sushma Rohatgi, Prof. Subir Sarkar, Prof. Indira Lepcha, Prof. Bijan Kumar Ghosh, Prof. Dipankar Bhattacharya, Prof. Swapan Kr. Bose, Prof. Arabinda Mukherjee, Dr. Jyotirmoy Sen and Dr. Neema D. Lama, Sri Aptar Sarkar, Sri Manish Chaklader, Sri Suresh Chandra Sarkar, Dr. Sumanta Chattoraj for their love, affection and encouragement.

Hearty thanks to the editorial assistance team specially Dr. Dipak Kumar Singh for providing tireless support in shaping the same. My special thanks to Mr. Gopal Das, my student for his cordial support. I am thankful to my dear friends especially Dr. Arindam Basak, Mr. Prasenjit Sarkar, Mr. Sanjoy Sarkar, Mr. Nayan Jyoti Datta and many unnamed well-wishers, colleagues and relatives who encouraged me lot in shaping the same. My gratitude to the contributors who have extended their hands to make the volume rich through contributing their valuable papers.

I wish to express my sincere thanks to my wife Mistu and our kids: Riddhima and Rutvik and my extended family members for their understanding, patience and sacrifice throughout this journey.

My heartfelt thanks to Mr. Rajesh Sharma and his entire team of Namya Publishing house for their utmost sincerity to publish this book in due time. Last but not the least I am indebted to almighty God for love and blessings.

Place: Siliguri

Date-16/07/2020

Dr. Ranjan Sarkar

Editor

www.drranjansarkar.in

Contents

Foreword *vii*

Preface *ix*

Acknowledgements *xi*

Part - I : Natural Environment and Sustainability

1. Excessive Rainfall and Resultant Urban Flooding, Associated Vulnerability and Mitigation: A Case Study of Alipurduar Town 3

   Tuhin Dey Roy & Dr. Arindam Basak

2. Spatio-Temporal Analysis of The Fluctuations of Groundwater Level: A Study in Rajasthan, India 18

   Parimita Saikia

3. A Case Study on Insect Biodiversity at Madarihat Youth Hostel and Beat Office Campus in Alipurduar District of West Bengal, India 29

   Mayukhmala Mandal

4. Present Scenario in Indian Sundarbans: A Biogeographical Overview 39

   Kantisree Goswami

5. Environmental and Social Impact of Super Cyclone Amphan 54

   Dr. Savita Mishra

6. Microclimatic Effect on Tea Cultivation: Rain Fall and Yield 69

   Dr. Ratula Mukherjee

Part - II : Environmental Challenges in the Present Context

7. Global Environmental Challenges in the 21st Century 81
Gopal Das & Dr. Ranjan Sarkar

8. Covid-19 Crisis Through A Migration Lens: Livelihood Effects for Pandemic Induced Displacement 102
Chandra Kanta Saha

9. The Problem of Immigration in Assam and Its Impact 118
Kanchan Devi

10. Environmental Degradation and its Impact on People of Bhubaneswar 128
Subhasmita Parida & Dr. Sudhir Kumar Kisku

11. River Pollution in India: A Socio-Legal and Sustainable Approach 148
Dr. Abhijit Bhattacharjee & Soumya Pratik Dutta

12. Covid-19 Pandemic in India: A Geo-Environmental Analysis 159
Biswajit Majumdar

13. Changing Paradigm in the Educational Environment During COVID-19 Pandemic: An Indian Perspective 168
Dr. Dipak Kumar Singh & Dr. Ranjan Sarkar

14. River Bank Erosion Induced Human Displacement and its Problems: A Study of Maynaguri Block in West Bengal, India 175
Dr. Madhusudan Karmakar

15. Bankline Migration of Nonoi River and its Impact on Human Life: A Case Study of Sipajahr Development Block, Darrang District, Assam, India 187
Dr. Sikha Rani Nath

16. A Study on Arsenic Contamination of Ground Water in Lalgola Block under Murshidabad District, West Bengal 199
Ananya Chowdhury

Part - III : Environmental Management and Practices

17. Application of Modern Geospatial Tools in Vegetation and Water Estimation: A Case Study of Burisuti Pathar Wetland Area of Assam — 219
Dr. Krishna Das

18. Cross-Disciplinary Approach Between Chemistry and Geography for Environmental Problems: An Overview — 233
Mominul Sinan

19. Interweaving Environment, Ecology and Feminism: Ecofeminism as The Means of Environment Conservation — 244
Swati Sharma

20. Estimation of Ground Water Resources of Chandernagore Municipal Corporation and Hugli-Chinsurah Municipality: A Comparative Analysis — 254
Dr. Malay Ganguli

21. A Perceptive Study on Tourism Practices and Environmental Quality with Respect to Tourism Footprints in Rishyap Village of Kalimpong District, West Bengal — 267
Dr. Anirban Roy & Partha Das

22. Water Management and Conservation in India — 284
Dr. Niraj Kumar

23. Impact of Nutrients in Our Society: An Analysis — 295
Keerthana. S.

24. An Assessment of the Status of Household Solid Waste and its Management: A Case Study of Jalpaiguri Municipality, West Bengal, India: An Insight — 312
Debarati Chowdhury & Dr. Indrajit Roy Chowdhury

List of Contributors — 343

Part - I

Natural Environment and Sustainability

1

Excessive Rainfall and Resultant Urban Flooding, Associated Vulnerability and Mitigation: A Case Study of Alipurduar Town

Tuhin Dey Roy[1] & Dr. Arindam Basak[2]

[1]Ph.D. Research Scholar & [2]Assistant Professor,
Department of Geography & Applied Geography,
University of North Bengal, Siliguri, West Bengal, India

Abstract

Floods can be defined as 'the submergence of usually dry area by a large amount of water that comes from sudden excessive rainfall, an overflowing river or lake, melting snow or an exceptionally high tide'. Urban Flooding is specific in the fact that it is caused by a lack of drainage in urban area. Increasing trend of urban flooding is a universal phenomenon and poses a great challenge to city administration and urban planners. "These (Alipurduar) areas are prone to heavy rains as they are close to the Bhutan hills. Sometimes the areas get over 200mm of rains in a day" (I.M.D). During the month of August in 2017 extensive area of Alipurduar Town submerged under water due to torrential rainfall for several days in continuation. Continuous dumping of garbage on river, sedimentation of river bed, lack of drainage facilities, degradation of

[1] Corresponding Author

wetland in town and some other anthropogenic causes make flood disaster a common phenomenon here. This flood has a widespread impact including temporary relocation of people, damage to civic amenities, deterioration of water quality and risk of epidemics etc. Proper Disaster Management Plan, City Drainage Plan, Master Plan/ CDP, Land Use Plan is the need of the hour to mitigates this problem.

Keywords: Alipurduar, Submergence, Urban Flooding, Degradation, Mitigation.

Introduction

Alipurduar Municipality is a Class II town in the newly formed district of Alipurduar in West Bengal. The total population of Alipurduar Municipality in 2011 was 65232 persons.

Although there are a number of urban centres in North Bengal with a population size higher than that of Alipurduar town but because of its location it occupies a very important position in this region. Located in the heartland of the tea plantation belt of the *Dooars* as well as its proximity with the neighbouring country of Bhutan and the neighbouring state of Assam has given it a special position of its own. Importance of Alipuirduar town has further increased manifold with the formation of Alipurduar district after bifurcation of Jalpaiguri district on 25th June 2014.

Alipurduar town is prone to high intensity rainfall as it is located close to the Bhutan hills. Sometimes the area gets over 200mm of rainfall within a day. During the month of August in 2017 extensive areas of Alipurduar Municipality submerged under water due to torrential rainfall which continued for several days.

Objectives

The basic objectives of this paper are:

1) To highlight the history of flood occurrences in Alipurduar town.
2) To find out the possible reasons for flooding in this town.
3) To evaluate the dangerous impact of flood on human lives.
4) To suggest some remedial measures to lessen its vulnerability.

Database and Methodology

This study has been completed following a combination of both empirical as well as well as descriptive method. In the study both primary as well as secondary sources of data has been used. Primary data related to various socio-economic parameters to analyze the impact of flood has been collected through questionnaire survey. Secondary information about the municipality has been collected from Census of India 2011 to supplement primary survey. Satellite images as a secondary source has assisted very much in describing the changes in river bed. Some mapping techniques and quantitative analysis have been used for the better comprehension of the scenario related to frequent flooding of the study area. Various statistical techniques, graphs, diagrams, maps and charts have been used to bring out meaningful co-relation among the parameters.

Table 1: Flood Occurrence in Alipurduar Town

Year	Rainfall in mm	Triggered by	Remarks
1993	850mm/ day	Cloud Burst over lower Bhutan- Darjeeling Himalaya	Town Area Was Devastated, Massive fluvial transformation taken place. This occurrence left a deep mark to the local people who still remember the aftermath and trauma.
1998	1250 mm/ 72 hours	Cloud Burst over lower Bhutan Himalaya	Flooded more than 2 days, Communication fully stopped.
2005	-	Because of heavy rainfall in the last week of August, Kaljani rives were flooded	People died, farmland affected
2012 25 June	226.20	Nonai	Waterlogging, death of a child, power cut
2013 7 July	244	Heavy Rainfall, River Responsible Kaljani, Nonai, Old Dima	Market area, 17 Out of 20 wards affected. Trees uprooted, transportation on derailed
2016 21 June & 25 July	175.4 & 330	Heavy Rainfall, River Responsible Kaljani	Affected Wards- 1,5,8,11,12,15,1 8,19. People Shifted, Schools shut,

Source: Disaster Management Plan, Alipurduar District & Chapter V: Flood Hazards in Jalpaiguri-...PDFshodhganga.inflibnet.ac.in>bitstream, Complied by Authors.

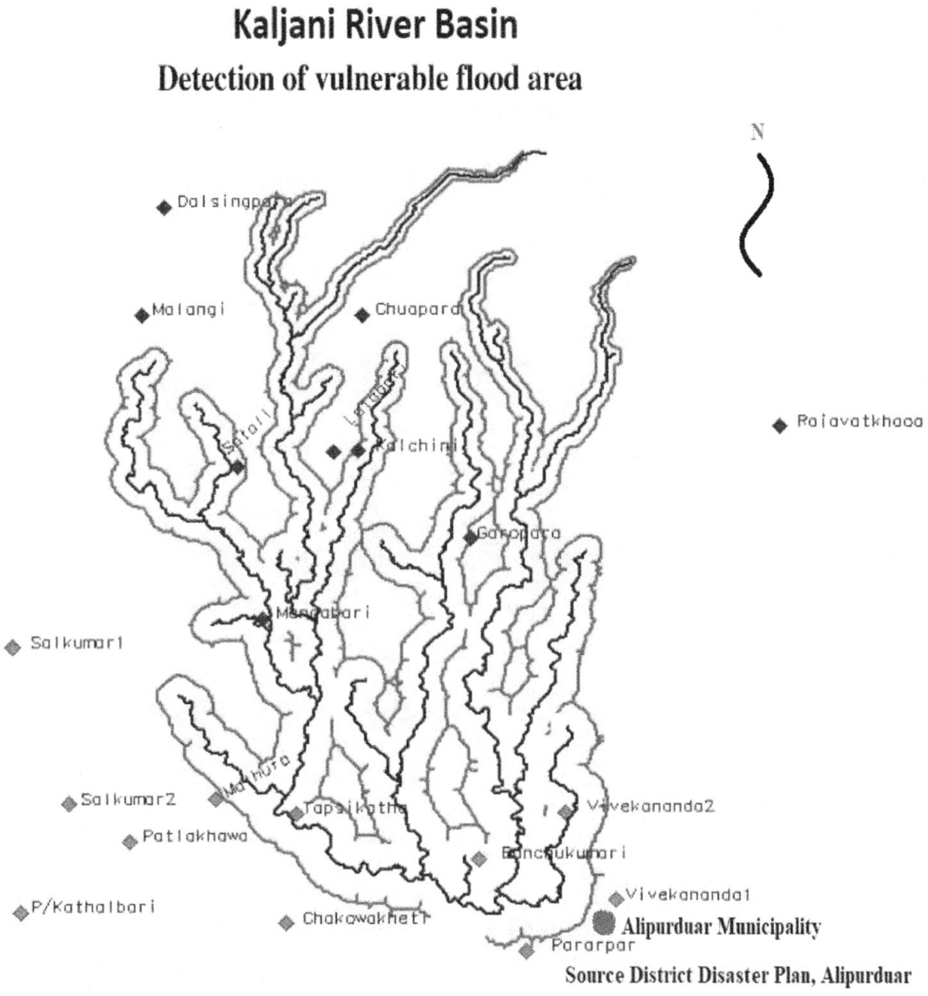

Figure 1: Kaljani River Basin

August 2017 Flood

According to the forecast, heavy shower was expected in the Dooars and Bhutan hills - where most Dooars rivers originate - in the next 24 hours in 8 August, 2017. A similar forecast was issued during 10 to 12th August 2017 and thousands of people in Binaguri, Banarhat, Jateswar in Jalpaiguri, Hamiltonganj, Coochbehar and Alipurduar were marooned under Water.

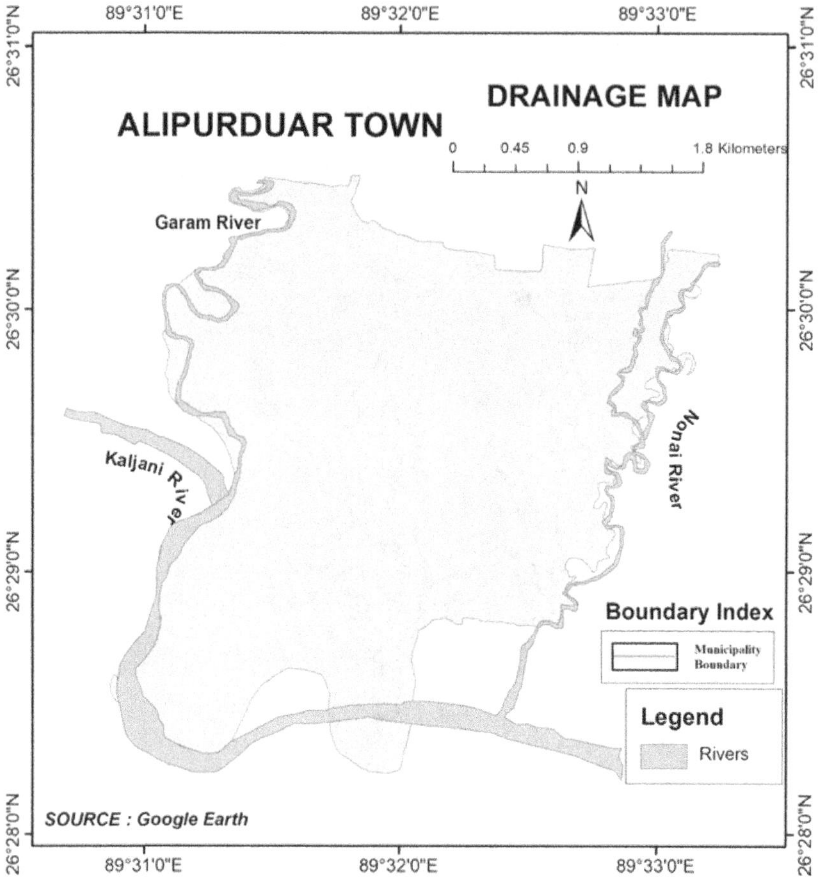

Figure 2: Drainage Map of Alipurduar Town

According to source of Irrigation department Banarhat in Jalpaiguri district received 335mm of rainfall, while Hashimara received 208mm of rainfall, Jalpaiguri recorded 129.4mm of rainfall, Nagrakata got 135mm rainfall and in Alipurduar the rainfall was 390.44 mm in the 12th August 2017. All the 20 wards of Alipurduar municipality were flooded. Some residents of wards 1, 5, 8, 11, 12, 15, 18 and 19 - all very close to Kaljani River have shifted to higher grounds. Tents have been erected for them at some places. Even the Alipurduar Chowpathi, New Alipurduar Station went under water at that time. Not only Mobile network has been suspended but also the power cut continued for two day. The Kaljani River was flowing above the extreme danger level. The flood affected people were short of food and drinking water.

A press release issued by NFR in 12.08.2017 said, 75730/75729 DEMU between Silghat Town-Guwahati, 15909 Awadh Assam Express between Dibrigarh-Lalgarh, 15660 Kanchanjungha Express between Agartala-Sealdah, 12378 Padatik Express between New Alipurduar-Sealdah and 13142 Teesta Torsa Express between New Alipurduar-Sealdah were cancelled.

Causes of Frequent Flood in Alipurduar

Gradual Increase of Rainfall

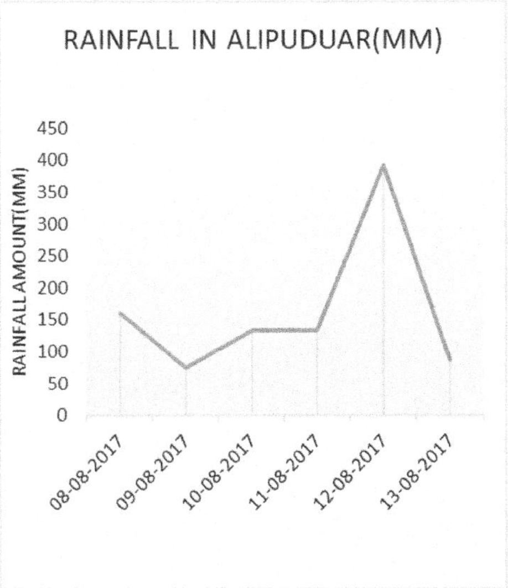

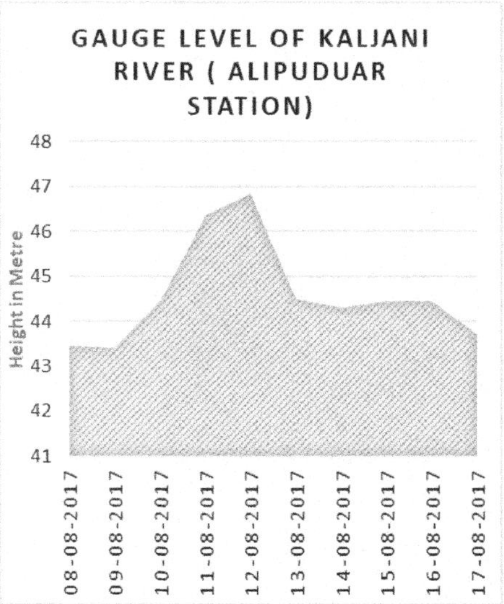

Figure 3 Figure 4

Source: http://www.wbiwd.gov.in/index.php/applications/dailyreport

We all know that Alipurduar District is the wettest district in West Bengal. Buxaduar in Alipurduar District has recorded the highest rainfall in West Bengal. In August 2017, the rainfall in this region finally increased. Through above mentioned the two graphs, we can give an overall picture of rainfall and river flow in this region.

If we correlate the amount of rainfall and the river gauge level, we can see that although there is a close relationship between the two in the first stage, but it took several days for the river to flow even though the amount of rainfall later decreased.

Table 2: Regression Analysis

Regression Model B		Unstandardized Coefficients		Standardized Coefficients	t	Sig.
		Std. Error	Beta			
1	(Constant)	11.690	9.438		1.239	.251
	Rainfall amount	.132	.062	.603	2.138	.065

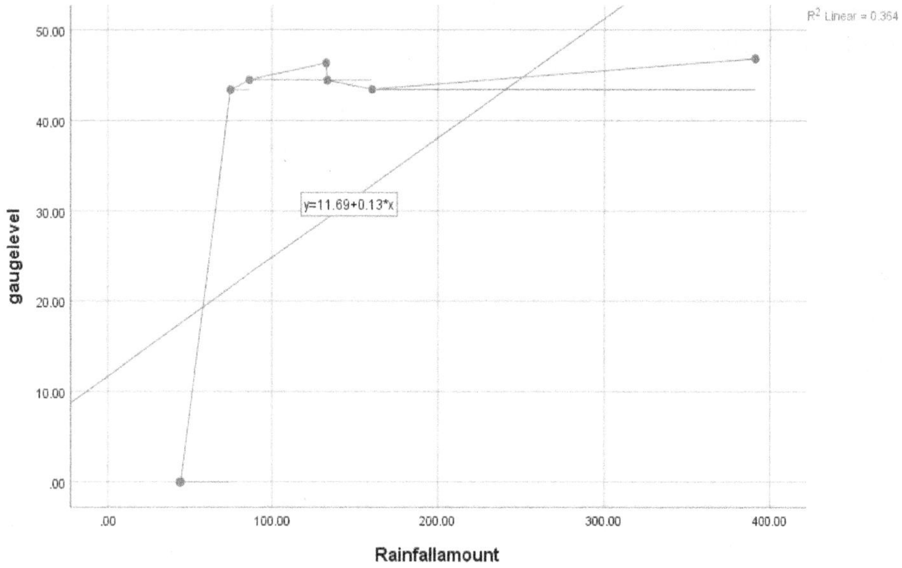

Figure 5

No Barrage on Bhutan Rivers

There are 72 rivers and streams which originate from Bhutan and flow through north Bengal (in the districts of Jalpaiguri, Alipurduar and Cooch Behar). There had been incessant rainfall in north Bengal but the flood was mainly caused by these rivers. There is no barrage in any of these rivers in Bhutan, as a result of which several areas in north Bengal are flooded. Govt. of West Bengal seeking for "India-Bhutan joint river commission" because with the view of state irrigation department "Formation of such a joint body comprising representatives of two countries can help in the exchange of information on flood and rainfall on a regular basis. Also, there can be regular discussions on the aversion of inundation in areas downstream the Bhutan rivers, that is north Bengal. We can place proposals like construction of barrages and water reservoirs in Bhutan and at some locations of north Bengal to regularise the flow of water. If the joint river commission is formed, it can largely help in resolving the problem."(Telegraph,2017)

Garbage Dumping on the River

Alipurduar is one of the cities of North Bengal. Alipurduar subdivision has been going on for 125 years, but no permanent arrangement has been made for garbage disposal in Alipurduar town till date. as everything from food scraps to plastic was dumped into the river In August 2017, the navigability of the river took a terrible turn.

Photo Plate 1

Sedimentation on Kaljani River

Bhutan does dolomite mining near rivers. Throughout the year and particularly during the monsoon, the rivers carry heaps of dolomite dust, boulders, pebbles and even trees and they get deposited on the riverbeds in north Bengal. Because of this, riverbeds become higher and banks are breached, inundating large swathes of land. This two-picture giving us a clearly idea about deposit on the land because of the inundation in the river bed near the town premises.

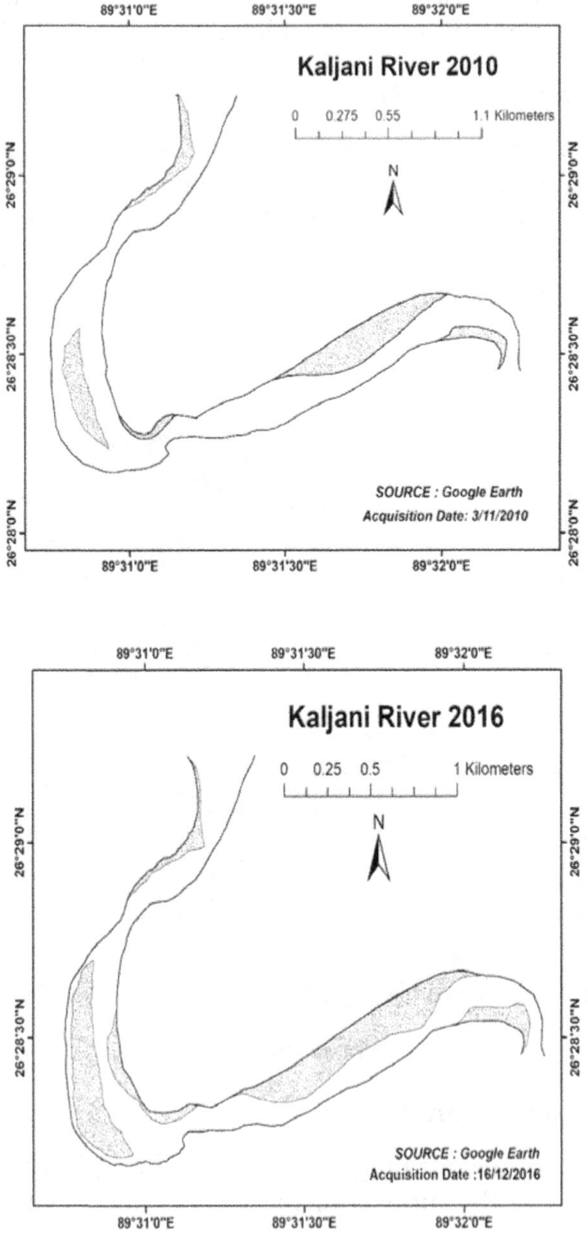

Figure 6: Sedimentation on Kaljani River

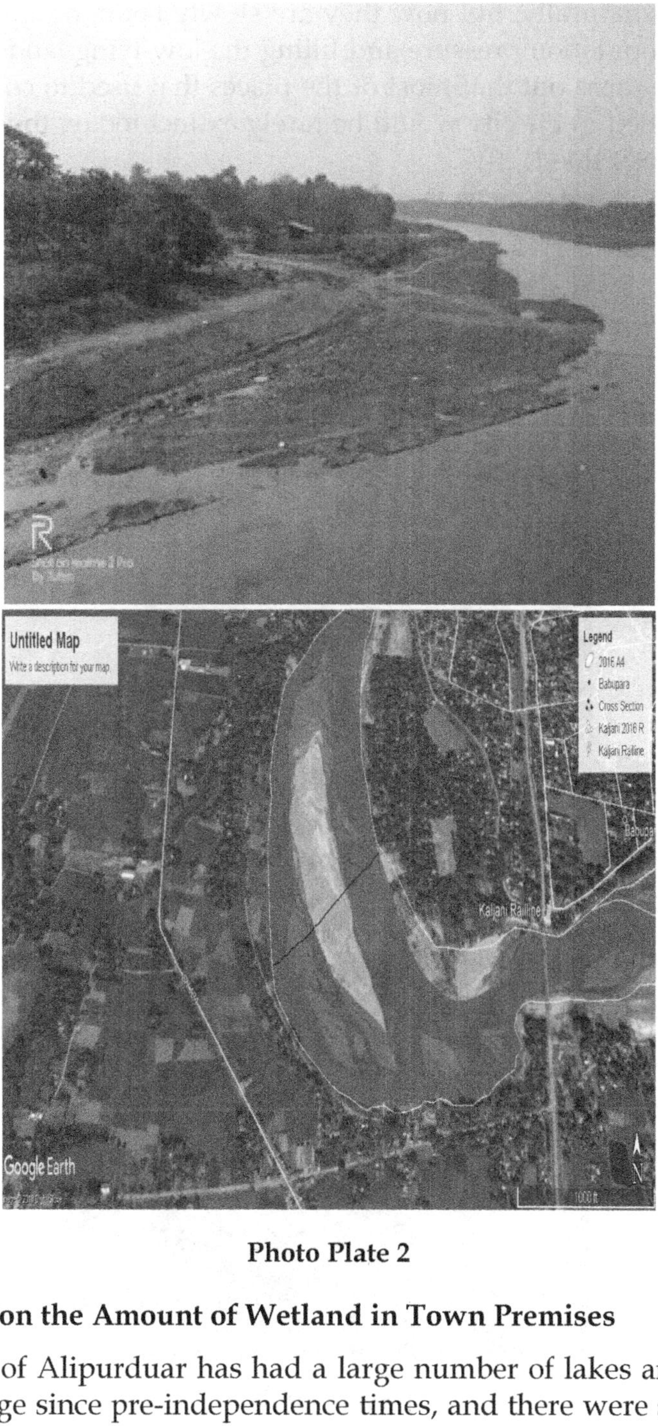

Photo Plate 2

Degradation the Amount of Wetland in Town Premises

The town of Alipurduar has had a large number of lakes and ditches for drainage since pre-independence times, and there were some low-lying lands naturally, but now they are slowly being occupied due to the huge population pressure and filling the low-lying lands is a huge problem. It turns out that most of the places that used to collect water when it rained in the city would be rarely extinct today, thus it turns a major cause of flood, 2017.

Photo Plate 3

Lack of Drainage Facilities

If we manage the city of Alipurduar according to District Disaster Management's plan, we will see that the city of Alipurduar still has inadequate drainage system in a manner of Sluice Gate and Embankment. All the wards along the river bank were severely inundated by the 2017 floods. Similar measures have not yet been implemented, resulting in the city overflowing into the Kaljani waters and flooding the surrounding by Nonai River.

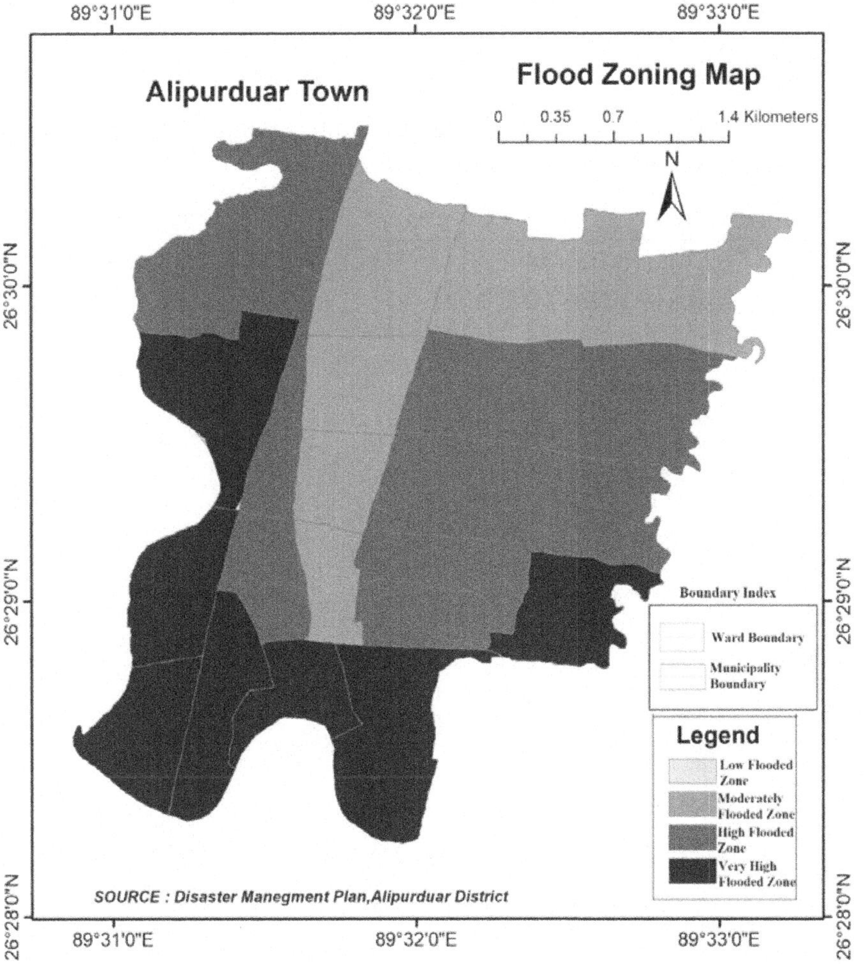

Figure 7: Flood Zoning Map of Alipurduar Town

Consequences on Environment and Human Lives

1) Some people has been died; Several Animals also died.
2) Many families have been marooned. A mass of people has been migrated.
3) There has been severe water crisis, water got contaminated. Acute water scarcity has been found in Alipurduar.
4) Roads have been devastated, waterlogged. Big cracks have been formed on the roads which have disturbed the entire transportation system.
5) Rail transportation has been stopped.
6) Agricultural crops have been damaged near town premises. For that very high price of daily needs.

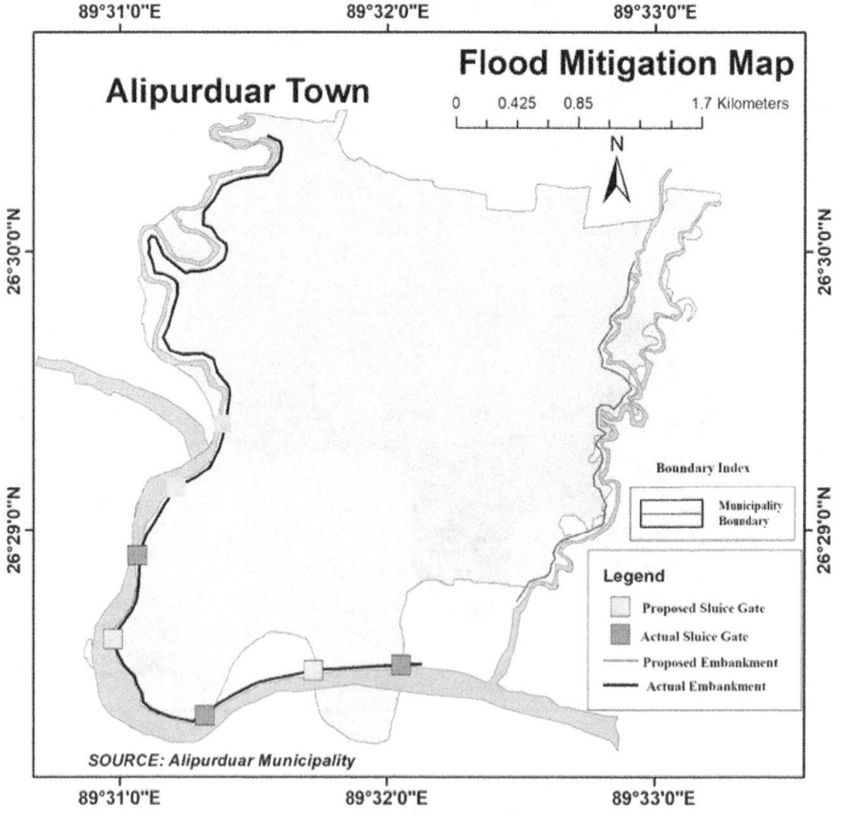

Figure 8: Flood Mitigation Map of Alipurduar Town

Conclusion

Thus, from our extensive study about the effect of flood on the people of Alipurduar Municipality it is very clear that the relationship between natural disasters and human society is a very intricate and complex one. The complexity further increases when human beings with their callous activities invite natural disasters like flood on their doorsteps. Our study area reflects the same story. Indiscriminate mining of sand and stones in the bed of Kaljani is gradually altering the channel morphology of the river which if continues for a long period of time will destabilize the dynamic equilibrium of the surrounding environment. The mitigation plan to minimize damage from flood may be as follows:

1) Make a Proper drainage plan For Town and Town Surrounding.
2) River Barrages should be made immediately on Bhutan rivers otherwise huge amount of sedimentation deposit in the low-lying rivers like Kaljani, Dima.
3) Bank protection measures should be undertaken by the Government of W.B in frequent Flood Prone Zone.
4) Garbage dumping ground or Municipal Solid Waste Plan must be set up in the Town.
5) Make continuous watching on flood venerable Embankment.
6) A Mass awareness in needed to make people conscious about after effects of unsanctified drainage system and disposal of materials in the river bed.
7) Flood forecasting is crucial to aware local inhabitants prior to the event and prevents such loss of human lives.

References

Barman, P. and Goswami, D.C., (2009) flood zone mapping of Guwahati municipal corporation area using Gis technology,10th ESRI India User Conference 2009, Geography in Action.

Bera et. al (2012). Application of Rs & Gis in flood management, a case study of Ongalkote blocks, Burdwan, West Bengal, India. International Journal of Scientific and Research Publications, 2 -11, 414-422.

Laha et.al. (2014), Flood hazard impact assessment on societal perspective, case study of lower catchment's of Kalighai Baghai and Kapaleswary river, International Multidisciplinary e-Journal, III-IV, 9-115.

Arnell, N. W., Clark, M. J., MGurnell, A. (1984), Flood insurance and extreme events: the role of crisis in prompting changes in British institutional response to flood hazard, Applied Geography, 4-2, 167-181.

O.N Dhar, S. Nandargi, A study of floods in Brahmaputra Basin In India, Int. Jour. Climatology, 20-21,771-781.

Pauleit, S. and Duhme, F. (2000) Assessing the environmental performance of land cover types for urban planning, Landscape and urban planning.,51- 52, 1-20.

Venkata, B. G and Sinha, R., (2005), GIS in Flood Hazard Mapping: A Case Study of Kosi River Basin, India, GIS Development: The Geospatial Resource Portal.

https://en.wikipedia.org/wiki/Alipurduar assessed on 8th February, 2018.

http://www.wbiwd.gov.in/pdf/ANNUAL_FLOOD_REPORT_2013.pdf assessed on 28h February, 2019.

http://www.imdpune.gov.in/Links/temp_rf_extreme.html assessed on 3rd March, 2018.

http://www.wbiwd.gov.in/index.php/applications/dailyreport assessed on 17th July, 2018.

http://www.ndma.gov.in/images/guidelines/flood.pdf assessed on 6th December, 2018.

http://www.apfm.info/publications/tools/APFM_Tool_14.pdf assessed on 3rd June, 2019.

http://alipurduar.gov.in/DMP2019.pdf 3rd July, 2020.Government of India Planning Commission Report of Working Group on Flood Management and Region-Specific Issues for XII Plan.

2

Spatio-Temporal Analysis of The Fluctuations of Groundwater Level: A Study in Rajasthan, India

Parimita Saikia
Ph.D. Research Scholar, Department of Geography,
The Assam Royal Global University,
Betkuchi, Guwahati, Assam India

Abstract

Groundwater is a major source of freshwater in many parts of the world. Some regions are excessively dependent on it leading to groundwater consumption faster than it can be naturally replenished and causing water tables to decline unremittingly. Rajasthan is the largest state in the country and water resources supporting the human population are very less. In addition to its water problems, Rajasthan is undergoing an industrial transformation and population growth in some of its districts. The objective of this study is to identify the groundwater critical zones, study the spatial behaviour of groundwater level fluctuations and analyse the trend of groundwater level during pre-monsoon and post-monsoon seasons in the state. Due to over-exploitation of exploitation of groundwater in dry regions can lead to irretrievable consequences and identification of critical areas can help sustainable use of groundwater. In this study, Inverse Distance Weightage (IDW) was applied for estimating the unknown attribute values of locations that are within the database using known data values. Results from the study illustratesthat critical zones are increasing spatially over the districts of Rajasthan due to population growth. Also, there has been decrease in groundwater level, i.e. maximum of 8.2 m bgl in the pre-monsoon season and 5.05 m bgl in the post-monsoon season between 1996 and 2015.

Keywords: Groundwater Level Depletion, Inverse Distance Weightage, Interpolation.

Introduction

Groundwater accounts for about 20 percent of the world's fresh water. But it is still less than 1 percent of all the water on Earth, comprising all ocean water and permanent ice. Groundwater refers to the water found beneath the Earth's surface. This water originates mostly from rain, melted snow, and other water that seeps through soil, or the cracks in the rocks. Due to gravity, the water moves downward, below the Earth's surface, until it hits a layer of rock or it can't get through, where it remains there and builds up. This section of water below the surface is called an aquifer. Groundwater can eventually go back up to the surface in the form of springs and wetlands (Chinnasamy et al., 2015).

In various places, groundwater discharged naturally into natural springs or contributed to rivers and wetlands. During droughts, groundwater often plays a vital role in sustaining rivers and streams and becomes a valuable buffer. Groundwater is a finite resource, and aquifers can become depleted when extraction rates exceed replenishment, or 'recharge', rates. Like surface water, groundwater can become polluted or contaminated (National Centre for Groundwater Research and Training).

Rajasthan is a state with low to extremely low rainfall, intense summers with very high temperatures, high diurnal variation of temperatures and low humidity and high evaporation. It is one such Indian state with complex agro-climatic zones and in urgent need of expanding groundwater resources. Moreover, increase in population and urbanization leads to groundwater depletion. Thus, groundwater study plays an important role in assessment, monitoring, planning, development and Integrated Water Resources Management in Rajasthan.

There are similar researches discussed in the literatures. Rede, 2012analyzed the spatio-temporal fluctuations of groundwater level to study depleting trend due to over-exploitation of underground water. Krishan et al, 2015 investigated the spatial distribution of groundwater levels on monthly basis in three districts of Northern Punjab in order to maintain sustainable usage of groundwater. Zhou et al., 2016 also attempted to study the spatio-temporal variation in groundwater level in Beijing. In this study factor analysis has been applied to identify the primary influencing factor in groundwater level depletion. Malla et al., 2014 and Sun et al. 2009 attempted to compare different interpolation methods in in groundwater study. Malla et al., 2014 compared Inverse Distance Weightage (IDW) and Natural Neighbor (NN) interpolation methods in groundwater quality. It was concluded from the study IDW as preferable spatial interpolation method.

The objectives of this study are to find the groundwater critical zones in Rajasthan, identify the seasonal variation of groundwater level and spatial trend of groundwater level fluctuation using inverse distance weightage for the years 1996, 2006 and 2015.

Materials and Methods

Study Area

Rajasthan is perceived as the 'Desert state' of India, widely has contrasting topography. Geographically, the state comprises from lofty hills from to sand dunes and fertile plains in the eastern side to arid areas towards its western and north-western part (Shukla, 2007, pp. 27-55). According to 2011 census, it has a total population of 68,548,437. The state accounts for more than 10% of India's geographical area, supports about 5% of the human population and 20% of the livestock but only possesses 1.2% of the total surface water and 1.7% of the groundwater available in India. The Aravalli hill ranges, running from north east to south-west, divide the state approximately into the western arid and eastern semi-arid regions. Rajasthan's economy has undergone considerable transformation in the recent past, with agriculture (including livestock) providing one-fourth of the state's GDP. Approximately 5.4 million households are engaged in farming, while 60% of the state's population depends on agriculture for their livelihood. Rajasthan is heavily dependent on groundwater for irrigation and about 90% of the drinking water and 60% of the irrigation water is sourced from groundwater supplies. The pressure on groundwater is further growing due to population growth and an increased number of industries. About 80% of the State areas have witnessed groundwater depletion and many towns and villages have experienced a shortage of drinking water, particularly in summer months (Chinnasamy et al., 2015).

Geographic Location and Physiography

The present study focuses on groundwater resources of Rajasthan and its districts (Fig.1). Rajasthan is the largest State in the country with total geographical area of 342,239 km^2 and comprises of 33 districts, 241 tehsils, 183 municipalities, 237 Panchayat Samities and 9,184 village panchayats (Yadav et al., 2016; Shukla, 2007, pp. 27-55). It is located between 13°3′ N to 30°12′ N latitude and between 69°30′E to 78°17′E longitude and shares its geographical boundaries with Punjab, Haryana, Uttar Pradesh, Madhya Pradesh and Gujarat.

The most striking geographical feature, Aravalli range- oldest folded mountain range in the world, which intersects the state diagonally end to end north-east to south west into three-fifth eastern desertic zone and two-fifth eastern semi-eastern region(Krishna et al., 2014).The state is divided into four physiographic units: a) North western and western desert or Thar desert, b) Aravalli Hills c) Eastern plains and d) South-eastern plateau. The physiographic units result in unique climatic set-up, soils and vegetation. The southern and eastern Rajasthan has relatively fertile areas and forms the Basin of Chambal river with its Banas, Kalisind and Parbati. The western region comprises of the Thar desert with only one river, arising southwest of Ajmer and joints the Rann of Kutch after covering a distance of 200 miles (Shukla, 2007, pp. 27-55).

Climate

The of Rajasthan varies from arid to sub-humid types. In the western part of Aravalli hills is characterized by low rainfall and erratic distribution, extreme diurnal and annual temperatures, low humidity and high wind velocity. The climate of eastern part of the Aravalli hills range from semi-arid to sub-humid. It has more or less same extremes of temperatures but comparatively lower wind velocity, high humidity with better rainfall (Shukla, 2007, pp. 27-55).

Soil and Vegetation

The soils of Rajasthan are complex due to diverse physiographic land features, range of distribution of rainfall and differing parent materials. Broadly, soils are divided into five groups: a) sandy soils b) sandy loam or light medium soils c) loam or medium soils d) clay loam or heavy soils and e) skeletal soils or shallow rocky and hilly soils. On the basis of the soil and climatic parameters, the vegetation of the state varies from dry tropical forests to desert thorn scrubs and grasslands (Shukla, 2007, pp. 27-55).

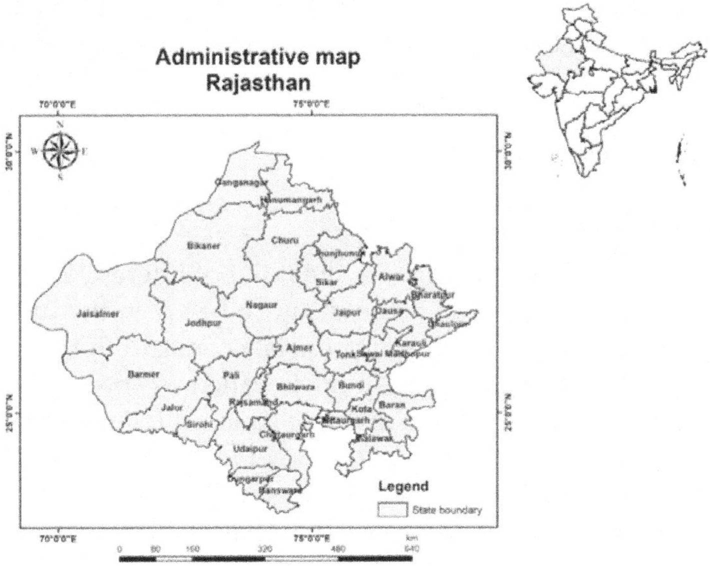

Figure 1: Administrative Map of Rajasthan

Data Collection and Processing

Data was obtained from India-WRIS WebGIS http://cgwb.gov.in/GW-data-access.html)which is a joint project of the Central Water Commission (CWC), Ministry of Water Resources, Govt. of India and Indian Space Research Organization (ISRO), Department of Space, Govt. of India to create a spatial database in MS-Excel for the years 1996, 2006 and 2015. The missing values were calculated by linear method using the formula given by:

$$\frac{a+(b-a)}{3} \quad \ldots(1)$$

where,

a = preceding year data value

b = succeeding year data value

Secondly, data was imported in ArcGIS software and Inverse Distance Weightage (IDW) was applied for estimating the attribute values of locations that are within the range of available data using known data values. In the inverse distance weightage interpolation method, the weights are solely a function of the distance between the point of interest and the sample points for i = 1, 2,.....n. Considering the distance D_i between these two points, a value of a point of interest point takes the form:

$$Z = \frac{\sum_{i=1}^{n} \frac{1}{(D_i)^q}}{\sum_{i=1}^{n} \frac{1}{(D_i)^q}}, \qquad \ldots(2)$$

where, Z is the interpolated value of the point of interest; Z_i is the value of sample point $i(i=1,\ldots,n)$; D_i is the distance between the interpolated and sampled values; and q is the appropriate constant. If the parameter q takes a value of 1 or 2, the method is called, respectively, inverse distance interpolation or inverse distance squared interpolation (Sun et al., 2009). In this study the value of q was set to 2. The area of Rajasthan was clipped from the interpolated data using Model Builder tool. The overall methodology adopted in the study has been shown in Figure 2.

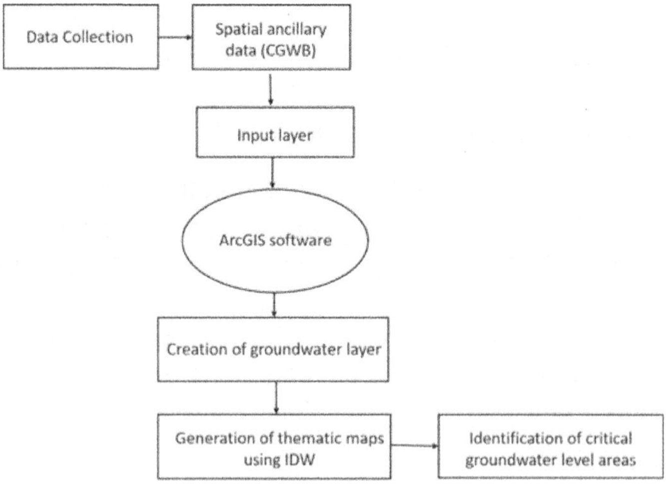

Figure 2: Methodology Adopted in The Study

Results and Discussion

There is a dynamic balance of water levels in aquifers between ground-water recharge, storage, and discharge. If recharge exceeds discharge, the volume of water in storage will increase and water levels will rise; if discharge exceeds recharge, the volume of water in storage will decrease and water levels will fall. It is because recharge and discharge are not distributed uniformly in space and time; ground-water levels are continuously rising or falling to adjust to the resulting imbalances.

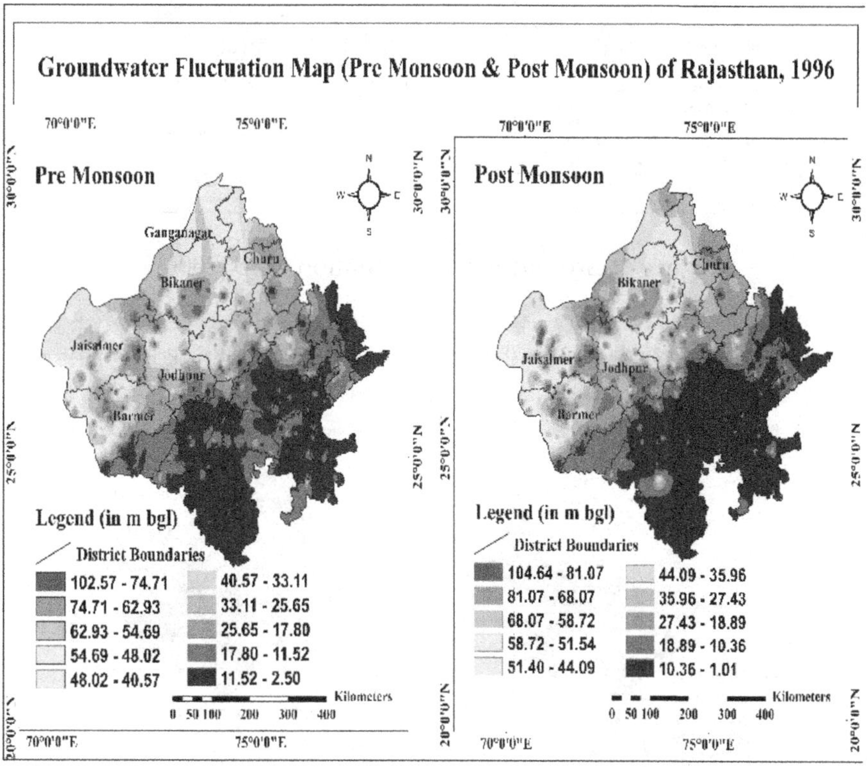

Figure 3: Groundwater Fluctuation Map (1996)

Figure 3 to Figure 5 illustrates groundwater level fluctuations for years 1996, 2006 and 2015 has been shown for pre-monsoon and post-monsoon seasons.

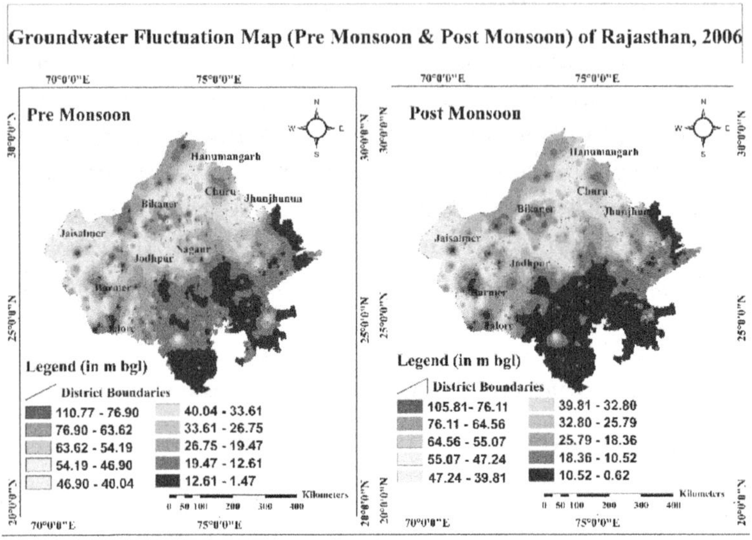

Figure 4: Groundwater Fluctuation Map (2006)

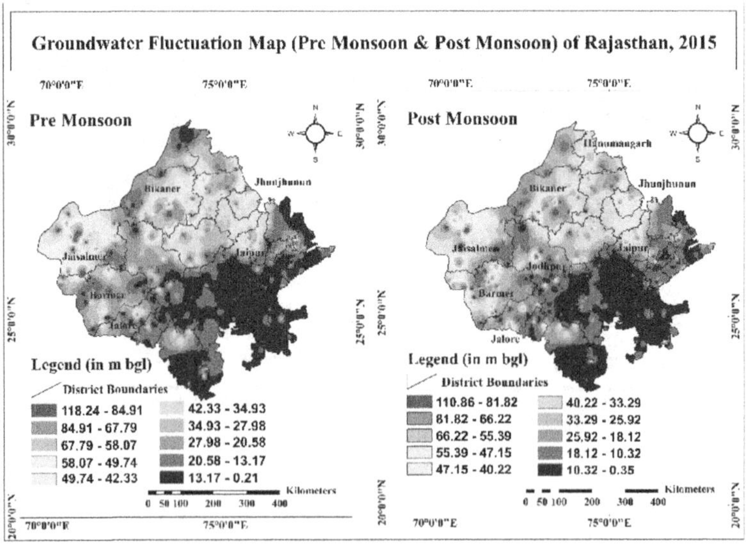

Figure 5: Groundwater Fluctuation Map (2015)

It can be proposed from the Fig. 3, in the year 1996, the groundwater level in post-monsoon was as low as 104.64 m bgl and as high as -1.01

m bgl. Figure 4 shows thatthe groundwater levels in the pre-monsoon was -110.77 m bgl and -105.81 m bgl in the post-monsoon season in the year 2006. Figure 5 illustrates the variation of groundwater level in 2015 seems to be more in the pre-monsoon season and the depth is as low as -118.24 m bgl and in the post-monsoon season, it is 110.50 m bgl. There has been decrease of groundwater level of 8.2 m bgl in the pre-monsoon season and 5.05 m bgl in the post-monsoon season between 1996 and 2015.

Conclusion

Water table in the state is continuously falling over the years. Over-exploitation in addition to the erratic distribution of rainfall has led to depletion of groundwater level at an alarming rate in spatially in the western and north-western part of the state. This study illustrates the trend of groundwater levels fluctuations in various network of observation wells districts of Rajasthan obtained from the Central Ground Water Board for both pre-monsoon and post-monsoon seasons. The figures show the fluctuations of groundwater level in Rajasthan for years 1996, 2006 and 2015. The critical zones have been observed majorly in the wells of Bikaner district over the years followed by Jaisalmer, Jodhpur, Barmer, Ganganager, Hanumangarh, Jhunjhunun, Jaipur and Jalore. However, the critical zones are increasing spatially over the districts of Rajasthan due to population growth. There has been decrease of groundwater level of maximum of 8.2 m bgl in the pre-monsoon season and 5.05 m bgl in the post-monsoon season between 1996 and 2015.Groundwater management is the major challenge in recent times. Therefore, the critical areas identified in the study should be focused in further studies for sustainable utilization of groundwater, which is the important driver of management of watersheds for sustainable development of the state.

References

Abdullahi, M. G., Toriman, M. E., Gasim, M. B., & Garba, I. (2015). Trends Analysis of Groundwater: Using Non-Parametric Methods in Terengganu Malaysia. *Journal of Earth Science and Climate Change, 6*(1). DOI:10.4172/2157-7617.1000251

Agarwal, E., Garg, R. D., & Srivastav, S. K. (2017). Spatio-temporal trend analysis of groundwater level in Unnao district, Uttar Pradesh, India.

International Journal of Creative Research Thoughts, 5(4), 1617-1626. Retrieved from: http://ijcrt.org/papers/IJCRT1704209.pdf

Chinnasamy P, Maheshwari B, Prathapar S. October 2015. Understanding Groundwater Storage Changes and Recharge in Rajasthan, India through Remote Sensing. *Water 2015*, 7, 5547-5565; DOI:10.3390/w7105547, 5548

Conlon, T. D., Wozniak, K. C., Woodcock, D., Herrera, N. B., Fisher, B. J., Morgan, D. S., &Hinkle, S. R. (2005). *Ground-water hydrology of the Willamette Basin, Oregon*. U. S. Geological Survey.Retrieved from: https://pubs.usgs.gov/sir/2005/5168/pdf/sir2005-5168.pdf

Krishan, G., Lohani, A.K., Rao, M.S, Kumar S., Takshi, K.S. (2015) Spatiotemporal Variability Analysis of Groundwater Level for Water ResourcesDevelopment and Management in Northern Punjab, India. J Environ Anal Toxicol 5: 279. DOI:10.4172/2161-0525.1000279

Krishna, P. H., Reddy, C. S., Meena, S. L., & Katewa, S. S. (2014). Pattern of Plant Species Diversity in Grasslands of Rajasthan, India. *Taiwania, 59*(2). DOI:10.6165/tai.2014.59.111

Kumar, S., Dhyani, B. L., & Singh, R. J. (2013). Depleting groundwater resources of Rajasthan state and its implications. Retrieved from:http://popularkheti.com/documents/2013-3/PK-1-3-14-64-68.pdf

Narjary, B., Kumar, S., Kamra, S.K., Bundela, D.S., Sharma, D.K. (2014). Impact of rainfall variability on groundwater resources and opportunities of artificial recharge structure to reduce its exploitation in fresh groundwater zones of Haryana. *Current Science*, Vol. 107, No. 8, 1305-1306.Retrieved from: https://www.currentscience.ac.in/cs/Volumes/107/08/1305.pdf

National Centre for Groundwater Research and Training: What is groundwater?. Flinders University GPO Box 2100, ADELAIDE SA 5001, ABN: 65 542 596 200. Retrieved from:http://www.groundwater.com.au/what-is-groundwater

Rede, H. N. (2012). A Study of Fluctuations in the Level of Underground Water in Jalna District, Maharashtra State, India. *Geoscience Research*, 3(2), 109. Retrieved from: https://bioinfopublication.org/files/articles/3_2_3_GEOR.pdf

Shukla, Y. (2007). *Monitoring Vegetation Degradation in Arid and Semi-Arid Regions of Rajasthan using SPOT-4 Satellite Data with particular reference to Surface Parameters*(Published doctoral disseration. Savitribai Phule Pune University, Pune). Retrieved from:http://lib.unipune.ac.in:8080/xmlui/bitstream/handle/123456789/122/11_chapter%202.pdf?sequence=11&isAllowed=y

Singh, A., Sharma, C. S., Jeyaseelan, A. T., & Chowdary, V. M. (2015). Spatio-temporal analysis of groundwater resources in Jalandhar district of

Punjab state, India. *Sustainable Water Resources Management, 1*(3), 293-304. DOI: https://doi.org/10.1007/s40899-015-0022-7

Sun, Y., Kang, S., Li, F., & Zhang, L. (2009). Comparison of interpolation methods for depth to groundwater and its temporal and spatial variations in the Minqin oasis of northwest China. *Environmental Modelling & Software, 24*(10), 1163-1170. DOI: https://doi.org/10.1016/j.envsoft.2009.03.009

Yadav SS, Kumar L, Dey A. January 2016. Ground Water Year Book 2014 - 2015, Rajasthan State. Government Of India, Ministry Of Water Resources, River Development & Ganga Rejuvenation, Central Ground Water Board, Western Region, Jaipur. Retrieved from:http://cgwb.gov.in/Regions/GW-year-Books/GWYB-2014-15 /GWYB%2014-15%20 Rajasthan.pdf

Zhou, Y., Xiao, W., Wang, J., Zhao, Y., Huang, Y., Tian, J., & Chen, Y. (2016). Evaluating spatiotemporal variation of groundwater depth/level in Beijing Plain, a groundwater-fed area from 2001 to 2010. *Advances in Meteorology, 2016*. DOI: https://doi.org/10.1155/2016/8714209

3

A Case Study on Insect Biodiversity at Madarihat Youth Hostel and Beat Office Campus in Alipurduar District of West Bengal, India

Mayukhmala Mandal
Assistant Professor, Department of Zoology,
Charuchandra College, Kolkata, West Bengal, India

Abstract

Insects are extremely diverse and important to ecosystems. The diversity and ecological importance of insects makes them very valuable for studies of biodiversity. Insects have great potential for understanding ecosystems and as measures of ecosystem health, but the incompleteness of knowledge and the limitation of resources increase the difficulty of work on insect biodiversity. The present survey was aimed to prepare a checklist of insect found in and around Madarihat Youth Hostel campus and Madarihat Beat office campus since there was no known published checklist of insect till date. Therefore, this is the first study which reports the abundance and distribution of insects in this area. Based on the data available, different order of insect- Orthoptera, Lepidoptera, Hemiptera, Colioptera and Diptera were found in the study area. Among these the most abundant insect order is Lepidoptera which was found almost in all sights and also highest in number. Overall biodiversity indices show relatively high biodiversity in the present campus area.

Keywords: Insect, Biodiversity Diversity Index, Species Abundance, Insect Conservation.

Introduction

Insects constitute a remarkably specious group of organisms attributed mainly to their small size, which allows them to occupy niches not available to larger organisms. They are adapted to highly differing environments from the Arctic tundra to Alpine mountain summits as well as warmer tropical rainforest and coastal mangrove swamps and are able to tolerate extremes of temperature and other environmental conditions. Insects are critical natural resources in ecosystems, particularly those of forests. In addition to their role as efficient pollinators and natural/biological pest control agents, some insect species are important indicators in ecosystems management. Insects are essential in the ecosystem by helping in nutrient recycling through leaf litter and wood degradation, carrion and dung disposal, and soil turnover. They play major role in plant pollination and maintenance of plant community composition. There is therefore the need for the conservation and protection of these species. Indeed, the forest canopy is the heart of biotic diversity and thus it is imperative that the forest be properly managed to ensure the persistence of such a unique group of nature, which is critical for ecosystem health. Understanding the extent of insect diversity is one of the major challenges in modern ecology. They have permeated the diverse and essential natural processes that sustain biological systems, making up over 75% of known species of animals. Now a day's India is occupying a significant space, documenting nearly 7 percent of global faunal diversity. There are about 7, 51,000 known species of insects, which is about three-fourth known species of plants and animals on the planet. The structure of insects has allowed them to exploit niches in almost all the habitats of the world. There are 1.4 million species of insects described in the scientific literature which is 80% of life currently recorded on earth. Insect is one of the five classes in the subphylum Uniramia. The Uniramia comprises those arthropods which have mandibles and pair of antennae, and other appendages which are primitively unbranched. The other four classes in the subphylum are collectively known as the myriapoda. 'Uniramia' occur worldwide and are predominantly terrestrial, but many of them live in freshwater and a few are marine representatives.

It is suggested that conservation of natural resources and biodiversity has become urgent issues in recent years for attaining an environmentally sustainable future. While a lack of data has historically excluded the use of many taxa as possible indicators. Therefore, careful targeting of any study is essential. Keeping in view the diverse and characteristic feature of insects,

Materials and Methods

Figure 1: Location of Study Area
(**Source:** https://goo.gl/maps/Xqwrro1SLQavSMyVA)

Study Area

The present study was conducted in Madarihat Youth Hostel Campus and the outskirts of Madarihat Beat office which is located at Latitude

E 089°16′31.8″ Longitude N 26°41′29.5″ respectively in the Alipurduar subdivision of West Bengal. Madarihat is a small town located at the fringe of Jaldapara National Park which is famous for Indian rhinoceros. It is located at the foothills of Eastern Himalayan along the bank of the river Torsa, at the national highway 31. This is one of the main tourist spots in North Bengal, surrounded by tea gardens, forest, hill and small rivers and also having a great diversity in the population.

Methodology

The study involved field visits to the entire campus from September 2018 to February 2019. The findings presented here are based on random survey. Soil sample were collected along with moist organic matter, wooden pieces, dry leaves and humus. An area measuring one sq. fit was measured with measuring tape, then soil is dug with the help of shovel, pushing it into soil and the soil was collected. The insects were collected from soil by Tullgren Method10PM to overnight and light trap method 11PM to overnight.

Data Analysis

The original data ware analyzed with the help of following index

1) Total Species. *S*- The number of species in each sample, that is, species with nonzero counts.
2) Total Individuals. *N*-The number of individuals each sample.
3) Frequency (%) =Total No. of quadrates studied) x100
4) Abundance = (Total No. of individuals of the species/ Total No. of quadrates in which the species has occurred)
5) Species Richness (Menhinick). $d = (S/N)$.

Result

This is the first study which reports the abundance and distribution of insects in the Madarihat Youth Hostel Campus, Alipurduar Jaldapara. Diversity study of the insect in the present campus can be utilized as an indicator of changing environmental conditions in the campus. In this study, a total of 70 individual of insects from 04 Orders were collected. They are from the order of Hymenoptera (Ant), Coleoptera (beetles), Acarina (mite), Collembola (spring tail).

Table 1: Study of Soil Sample by Tullgren Method at Madarihat Youth Hostel Campus

Specimen	Order	Number in the collected sample		Total number found in sample
		Site A	Site B	
Ant	Hymenoptera	15	9	24
Springtail	Collembola	-	6	6
Mite	Acarina	6	12	18
Beetle	Coleoptera	12	10	22

Source: Computed by researcher from the field study collected from Sept.,2018 to Feb.,2019.

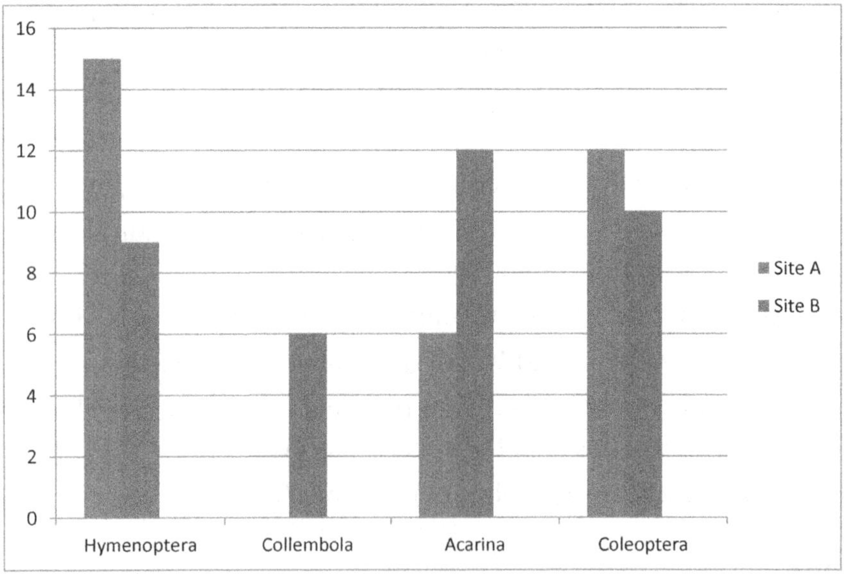

Source: Computed by researcher from the field study collected from Sept.,2018 to Feb., 2019.

Figure 2: Soil Sample by Tullgren Method at Madarihat Youth Hostel Campus

Table 2: Frequency, Abundance, Species Richness Index of The Soil Micro Arthropod in Madarihat Youth Hostel Campus Followed by Tullgren Method

Specimen	Frequency (%)	Abundance	Species Richness Index	
			Site A	Site B
Ant	100	12	0.522	0.658
Springtail	50	3		
Mite	100	9		
Beetle	100	11		

Source: Computed by researcher from the field study collected from Sept., 2018 to Feb., 2019.

In this study, a total number of 66 individual of insects from 05 Orders were collected from the outskirts area of Madariahat Beat Office by Light Trap method. They are from the order of Orthoptera (Grasshoper & Crickets), Lepidoptera (Moth & Butterfly), Hemiptera (Plant bug), Coleoptera (beetles) and Dioptera (Beetle).

Table 3: Collection and Identification of Specimen by Lighted Trap Method at Madarihat Beat Office

Specimen	Order	Number in the collected sample			Total number found in sample
		Site A	Site B	Site C	
Grasshoper & Crickets	Orthoptera	05	-	-	05
Moth & Butterfly	Lepidoptera	10	10	05	25
Plant bug	Hemiptera	04	06	04	14
Beetle	Coleoptera	05	06	04	15
Flies	Dioptera	04	-	03	07

Source: Computed by researcher from the field study collected from Sept.,2018 to Feb., 2019.

Table 4: Frequency, Abundance, Species Richness Index of the different Insect in Madarihat Beat Office followed by Light Trap method

Specimen	Frequency (%)	Abundance	Species richness index		
			Site A	Site B	Site C
Grass hoper & Crickets	33.33	05	0.945	0.639	1
Moth & Butterfly	100	8.333			
Plant bug	100	4.666			
Beetle	100	05			
Flies	66.67	3.5			

Source: Computed by researcher from the field study collected from Sept., 2018 to Feb., 2019.

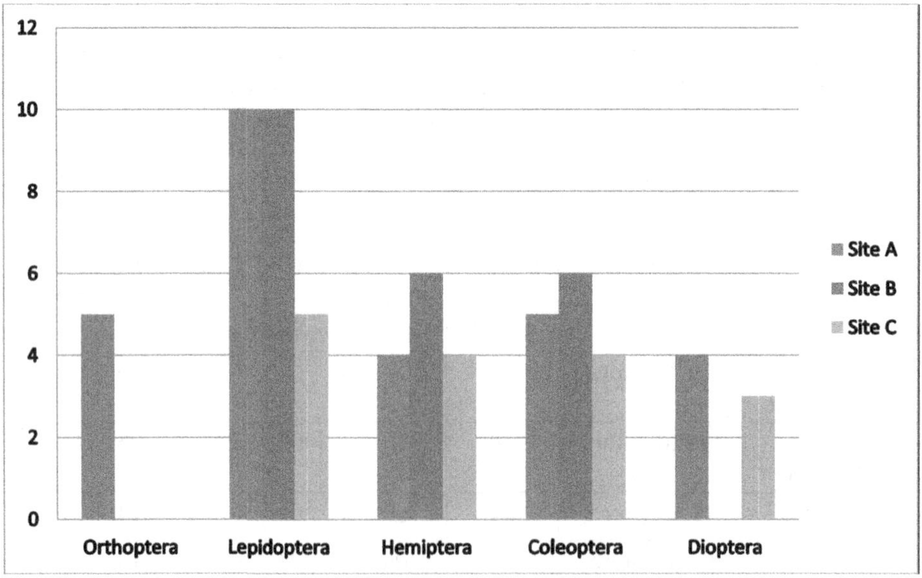

Source: Computed by researcher from the field study collected from Sept., 2018 to Feb., 2019

Figure 3: Identification of Specimen by Lighted Trap Method at Madarihat Beat Office

Discussion

From the **Tullgren** study **of Madarihat Youth Hostel Campus** Arthropods like Ant and Beetle ware found almost same in number. These data clearly suggest that the collect soil from quadrate sampling neither be dominated by any specific arthropod species. This clearly suggests that the area is highly biodiversity prone.

Analyzing the frequency and Abundance data from Madarihat Youth Hostel Campus, the Ant species was observed to have both high in frequency and abundance. Although mite and beetle have same frequencies of ant species, but their abundance value is lower than ant species. However, the springtail which is found to be common in that area but was not found in the sampling quadrate, (Frequency = 0, Abundance = 0).The frequency value is almost same in case of ant, mite and beetle and the abundance value of Ant, mite and Beetle.Therefore, it reflects that both quadrate have same species riches, i.e,also being observed when be calculate the species richness as it was found to be almost similar, (S-1= **0.522**, S-2= **0.658**). For this reason, Site-2> Site-1 according to species availability.

Analyzing the frequency and Abundance data from **Madarihat Beat Office by light trap method** the most abundant Insect order is Lepidoptera, which was found almost in all sight of light trap and also highest in number in frequency and abundance. According to species richness it was found that S-1=0.945, S-2=0.639 and S-3=1.For this reasonSite-3>Site-1>Site-2, according to species availability.

The biodiversity (diversity index, species abundance) of insect fauna in the present campus is mainly due to the rich vegetation in this area as vegetation plays an important role for the existence of insect fauna in a community as it provides the main source of food etc. for insects. This diversity has been modified at times and has tried to sustain itself in changing circumstances. The outcome of the study can be used constructively in planning sustainability of both man and natural environment. The various insect orders that were identified during the brief study have only highlighted the potential magnitude of biodiversity on the campus. The estimated numbers insects on the Campus are the products of the mosaics of critical habitats and the corridors that link them. Thus, it is important to maintain this connectivity in order to sustain the rich insect biodiversity on the campus, a system of monitoring the critical habitats and indicator species has to be evolved. Overall, our results highlighted that a small compact area like a segregated Youth hostel and Beat office campus supports a diverse insect species.

Conclusion

This study was an attempt to analyze some aspects of biodiversity of insects from **Madarihat Youth Hostel Campus** and the out skirts area of **Madarihat Beat Office**. Since it is a preliminary study, a lot of research is necessary in this regard and further collections are essential for getting a detailed record of the faunal diversity of insects and development of standard monitoring procedures for assessing the environmental stability in this area.

References

Belamkar, N. V., Jadesh, M. A. 2014. Preliminary study on abundance and diversity of insect fauna in Gulbarga District, Karnataka, India, IJSR. 3(12), 1670-1675.

Bindulekha D. S. Amalnath. S.2016 A Preliminary Study on the Biodiversity of Insects Collected from A College Campus: Thiruvananthapuram District, Southern Kerala IJSRISSN (Online): 2319-7064.

Choudhary, A., Ahi, J. 2015. A review: Biodiversity of fresh water insects, Int. J. Eng. Sci. 4(10), 25-31.

Danks, H.V. 1996. How to assess insect biodiversity without wasting your time, Biological Survey of Canada (Terrestrial Arthropods), 22 pp.

Finnamore, A.T. 1996. The advantages of using arthropods in ecosystem management. A brief from the Biological Survey of Canada (Terrestrial Arthropods), 11 pp.

Kulshrestha, R., Jain, N. 2016.A note on the biodiversity of insects collected from a college campus of Jhalawar district, Rajastan, Biosci. Biotech. Res. Comm. 9(2), 327- 330.

M. Ragaei, M. Allam Insect conservation and diversity a Journal of Islamic Academy of Sciences 10:2, 43-48, 1997.

P. Jugular and P. S. Cranston, Insects: An Outline of Entomology, Chapman and Hall, 2010.

Rosina Kyerematen, Daniel Acquah-Lamptey, Erasmus Henaku Owusu, 1 Roger Sigismund Anderson, and Yaa Ntiamoa-Baidu. 2014Insect Diversity of the Muni- Pomadze Ramsar Site: An Important Site for Biodiversity Conservation in GhanaJournal of Insects, 11 pages.

Samways, M.J. 1994. Insect Conservation. EOLSS Publications.

Spellerberg, I. F., Fedor, P.J. 2003. A tribute to ClaudeShannon (1916–2001) and a plea for more rigorous use. of species richness, species diversity and the "Shannon-Wiener" Index. Global Eco. Biogeo. 12(3), 177-179.

Wiggins, G.B. 1983. Entomology and Society, Bullet. Ent. Soc. America. 29, 27-29.

4

Present Scenario in Indian Sundarbans: A Biogeographical Overview

Kantisree Goswami

Assistant Professor, Department of Zoology,
Charuchandra College, Kolkata, West Bengal, India

Abstract

The Sundarbans is a mangrove area in the delta formed by the confluence of the Ganges, Brahmaputra and Meghna Rivers in the Bay of Bengal. It spans from the Hooghly River in India's state of West Bengal to the Baleswar River in Bangladesh. It comprises closed and open mangrove forests, agriculturally used land, mudflats and barren land, and is intersected by multiple tidal streams and channels. Four protected areas in the Sundarbans are enlisted as UNESCO World Heritage Sites, viz. Sundarbans National Park, Sundarbans West, Sundarbans South and Sundarbans East Wildlife Sanctuaries.

The Sundarbans mangrove forest covers an area of about 10,000 km^2 (3,900 sq mi), of which forests in Bangladesh's Khulna Division extend over 6,017 km^2 (2,323 sq mi) and in West Bengal, they extend over 4,260 km^2 (1,640 sq mi) across the South 24 Parganas and North 24 Parganas districts. The most abundant tree species are sundari (*Heritiera fomes*) and gewa (*Excoecaria agallocha*). The forests provide habitat to 453 faunal wildlife, including 290 bird, 120 fish, 42 mammal, 35 reptile and eight amphibian species.

Despite a total ban on all killing or capture of wildlife other than fish and some invertebrates, it appears that there is a consistent pattern of depleted biodiversity or loss of species in the 20th century, and that the ecological quality of the forest is declining. The Directorate of Forest is responsible for the administration and management of Sundarban National Park in West Bengal. In Bangladesh, a Forest Circle was created in 1993 to preserve the forest, and Chief Conservators of Forests have been posted since. Despite preservation commitments from both Governments, the Sunderbans are under threat from both natural and human-made causes. In 2007, the landfall of Cyclone Sidr damaged around 40% of the Sundarbans. The forest is also suffering from increased salinity due to rising sea levels and reduced freshwater supply. Again in May 2009 Cyclone Aila , and in May 2020, the super-cyclone Amphan have devastated Sundarbans with massive casualties. Securing the future of the Sundarbans, its biodiversity and people requires a long-term vision that can integrate climate adaptation and conservation strategies along with shorter term interventions such as ensuring sustainable livelihoods, access to clean and sustainable energy and effective human wildlife conflict management.

Keywords: Sundarbans, World Heritage Site, Mangroves, Biodiversity loss, Cyclones.

Introduction

The Sundarban, covering about one million ha in the delta of the rivers Ganga, Brahmaputra and Meghna is shared between Bangladesh (~60 %) and India (~40 %), and is the world's largest coastal wetland. The Sundarban (21° 30′ to 22° 40′ N, 88° 05′ to 89° 55′ E) comprises essentially of numerous islands formed by the sediments deposited by three major rivers, the Ganga, Brahmaputra and the Meghna, and a dense network of smaller rivers, channels and creeks. Sunderban is an enormous tract of compact mangrove forest in the world. Its importance for its changing floristic composition, economic use and wild life habitat has created a unique biotic region in the vast areas of Sundarbans. People are dependent on Sundarbans directly or indirectly for their livelihood. Government also earns revenue from the forest products such as timber, fuel wood, pulpwood, matchwood, thatching materials, honey, bee wax etc. Fishermen catch brackish water fishes of high commercial value. Common people capture bagda fry, which yield them at present the socio-economic stability. Biotic Region is a region of prescribed areas or physical habitat characterized with individual and popular components of living biota of the eco system. Located in the delta of the three rivers, Ganges, Brahmaputra and Meghna, it is home to a significant portion of one of the world's largest contiguous block of mangrove forests and biodiversity. In 1987, the Sundarbans was designated as a UNESCO world heritage site and thereafter was declared as a protected biosphere reserve by the Government of India in 1989. The Sundarbans display high biodiversity as well as the occurrence of endangered and highly threatened species, including the only population of critically endangered Royal Bengal Tigers (*Panthera tigris tigris*) found in a coastal mangrove habitat. Sundarbans is also home to 300 species of flora and about 425 species of wildlife, including the Gangetic Dolphin (*Platanista gangetica*) and Irrawady Dolphin (*Orcaella brevirostris*). The Sundarbans account for 85 per cent of all mangrove habitats found in India; including 63 of the 69 mangrove plant species found in the country. Flora and fauna of the mangrove forest enrich this biotic region of Sunderbans to such an extent that Government of India declared the entire 9630 sq. km of Sunderban as the Sunderban Biosphere Reserve in 1989 (Fig 1).

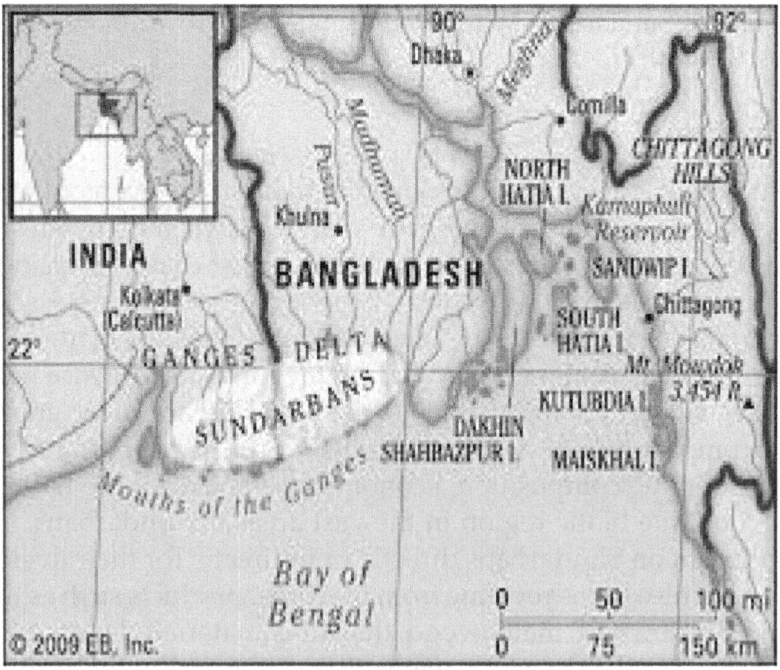

Source: Taken from Encyclopaedia Britannica

Figure 1: Sundarbans Geographical Region, Asia

Bio-diversity of the mangrove's ecosystem including algae, fungi, mangroves and tidal marshes, prawn (*P. monodon*) and occurrences of its seeds, Royal Bengal Tiger and estuarine crocodiles make the Sunderbans an important biotic region. But after years of abuse from man and nature, the mangroves seem to be nearing their limits. Illicit logging, mostly for building materials to house the region's booming population, has thinned out the periphery of the forest. At the same time, increasing water salinity caused by the encroaching sea is killing off many higher value, storm-stopping tree species, such as the sundari that gives the forest its name. The salinity assault comes from both land and sea: Upstream dams on rivers in India have reduced freshwater flow into the Sundarbans, while sea-level rise caused by climate change is flushing more salt water into the mangroves. Frequent cyclonic storms have also affected the area severely.

Due to such geophysical, climatic and most importantly, anthropological factors, the overall biological diversity of Sundarbans is being significantly lost.

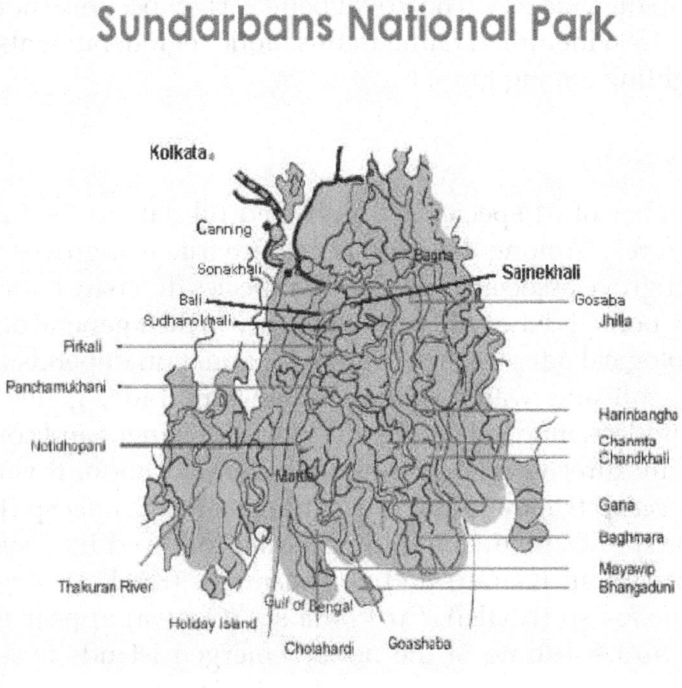

Source: from sundarbansnationalpark.com

Figure 2: Sundarbans National Park Map

Biological Diversity of Sundarbans: An Overview

Sundarbans is famous for its rich biological diversity. Enormous amounts of sediments carried by the rivers contribute to its expansion and dynamics. Salinity gradients change over a wide range of spatial and temporal scales. The biodiversity includes about 350 species of vascular plants, 250 fishes and 300 birds, besides numerous species of phytoplankton, fungi, bacteria, zooplankton, benthic invertebrates, molluscs, reptiles, amphibians and mammals. Species composition and community structure vary east to west, and along the hydrological and salinity gradients. Sundarban is the habitat of many rare and endangered animals (*Batagur baska, Pelochelys bibroni, Chelonia mydas*), especially the Royal Bengal tiger (*Panthera tigris*). There are 64 species of mangroves and their associated species and 1586 species of fauna are identified till date in the biotic region of Sunderbans. Mangroves provides habitat for many forms of both aquatic and terrestrial animals. Mangroves not only dominate the habitat and characterize the ecosystem but also yield economic resource such as capture of bagda fry is of commercially important to the villagers. The Royal Bengal Tiger occupies not only the pinnacle of 1586 identified fauna in this biotic region, but it also creates thrill of sighting among tourists.

Floral Diversity

A total number of 64 species are identified till date in the Sunderban mangrove forest. Among these 34 species are true mangroves while the rest are mangrove associated plants. The species diversity of mangroves is relatively poor and they all show similarity in their general occurrence and physiological adaptations. Mangrove zonation depends upon soil characters, soil and water salinity, tidal amplitude, gentle sloping, shallow mudflats, mud substrates etc. Floristic changes and community structures are directly related to the mudflat elevation. It is observed that *Porterasia* sp (Dhani grass) and saplings of *Avicennia* sp (Baen) are the pioneer species in an emerging mudflat followed by *Sonneratia* sp (Keora), *Ceriops* sp (Garan) and *Bruguiera* sp. (Kankra), *Aegialitis* sp (Tara), *Aegiceros* sp (Khalsi), *Excoecaria* sp. (Genwa) appear thereafter when the mud substrate of the newly emerged islands is stable and the top of the island is no more inundated. Mangrove associated palm species are found to grow on the side of 31 large and numerous narrow creeks and tidal inlets around the 48 mangroves dominated islands out of total 102 islands of Sunderbans. Mangrove vegetation of intertidal mudflat enhances the stability of the newly built up island sand also helps in configuring the new landmass.

Mangrove ecosystem has the significance as the coastal stabilizers, shelterbelt areas, nursery grounds for aquaculture, production of tannin, timber, paper and pulp, charcoal and several other by-products. Mangroves enhance land accretion, stabilize embankment and act as the green wall of the sea as they provide protection from the extensive damages caused by the waves and tsunamis. Mangrove forest known other way as 'Tidal forest' or 'Coastal wetland' has the ecological, economic and social significance. Mangrove ecosystems maintain the most important links between the land and the sea particularly along the deltaic estuaries.

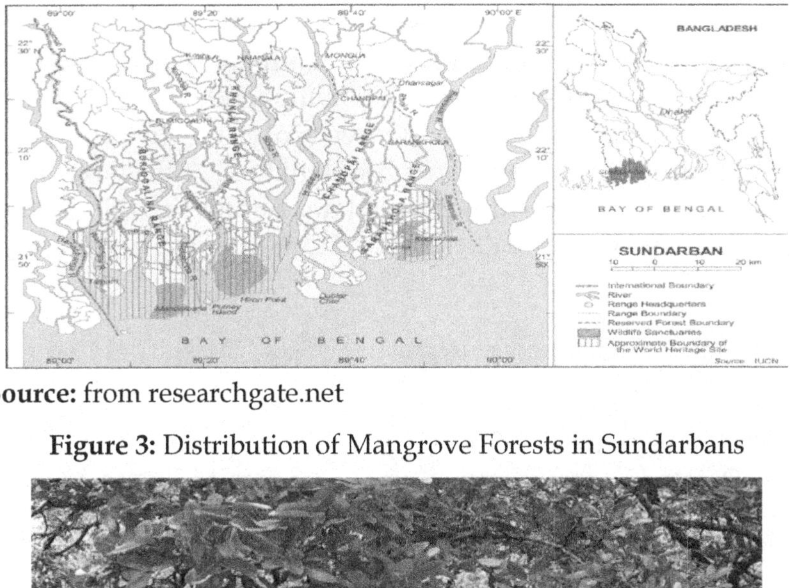

Source: from researchgate.net

Figure 3: Distribution of Mangrove Forests in Sundarbans

Source: from dhakatribune.com

Figure 4: Mangroves of Sundarbans

Faunal Diversity

Among numerous and rich faunal diversity of a large number of 1586 species, the tiger occupies the pinnacle of the mangrove ecosystem. The famous estuarine crocodiles of Sundarbans are now declared endangered species. Of late the occurrences of king cobras and Indian rock python are recorded around buffer areas of Sunderbans. A number of species among mammals is recorded endangered namely Smooth Indian otter, Gangetic dolphin and Irrawadi dolphin, Fishing cats and Leopard cats, Black porpoise etc. The fishes in estuaries of Sunderbans among fin fishes include a variety of 120 species. There are abundant occurrences of tiger shark, dog shark and 22 species of prawn in the Sunderban river waters. Very common birds in the Sunderbans are predator birds like white-belliedsea eagle, osprey, fishing eagle, bramhani kites, monsoon herons like open bill stork etc. Some migratory birds from Siberian region like curlew, whimbrel, plover, goliath and heron also breed in the mangrove forest of Sunderbans. Phytoplankton and zooplankton is important constituent in the mangrove ecosystem. In Sunderbans, 36 (species) of phytoplankton and 59 zooplankton, 22 Bacteria, 16 algae, 184 fungi species are identified till date. Sunderban biodiversity is enriched with the innumerable biota of 1586 species from protozoa to mammals.

Source: the independentbd.com

Figure 5: Faunal Diversity of Sundarbans

Loss of Biodiversity: The Visible Trend

Though Indian Sundarbans support huge biological diversity, but, unfortunately, it is being diminished significantly in the 21st century, due to several reasons. Firstly, this area has been frequently affected by several cyclonic storms in last few years, which have badly destroyed the natural ecosystem. Moreover, anthropological causes are also there. Increasing urbanization has considerably affected the biodiversity loss. Javan rhino, wild buffalo, hog deer, and barking deer are now extinct from the area. Large areas of the Sundarban mangroves have been converted into paddy fields over the past two centuries, and more recently into shrimp farms. The Sundarban has been extensively exploited for timber, fish, prawns and fodder. The regulation of river flows by a series of dams, barrages and embankments for diverting water upstream for various human needs and for flood control has caused large reduction in freshwater inflow and seriously affected the biodiversity because of an increase in salinity and changes in sedimentation. *Heritiera fomes* (locally called Sundari, from which Sundarban derives its name), *Nypa fruticans* and *Phoenix paludosa* are declining rapidly.

Loss of forest cover over the years AREAS IN SQ.KM.

Island	1986	1996	2004	2012
Gosaba	517.47	517.44	506.71	506.69
Dulibhasani West	180.03	174.78	170.90	163.47
Dalhousie	76.60	72.39	68.09	64.24
Bhangaduni	40.44	35.15	29.55	24.91
Jambudwip	6.09	5.91	5.68	5.003

SOURCE: MANGROVE FOREST COVER CHANGES IN INDIAN SUNDARBAN (1986-2012) USING REMOTE SENSING AND GIS, SCHOOL OF OCEANOGRAPHIC STUDIES, JADAVPUR UNIVERSITY

Source: thehindu.com

Figure 6: Impact of Climate Change in Sundarbans

Common Threats

A. Increasing Temperatures

Between 1980 and 2007, it has been observed that the temperature of the waters in the Sundarbans has increased at an accelerated rate of 0.5°C per decade compared to the observed global sea surface temperature warming at the rate of 0.06°C per decade. This accelerated increase in temperature of the sea has severe implications on aquatic life. This change greatly impacts the Sundarbans area as it is an estuarine delta. Further, it detrimentally affects the health of the mangrove ecosystem.

B. Rising Sea-Levels

In the past 25 years, sea level has raised at a rate almost double the global average. This is due to a combination of factors including land subsistence patterns. Due to continuous submergence in higher water, as an implication of rise of sea level, the plants are being noted to be shorter and narrower with fewer branches and leaves resulting in lower rates of photosynthesis and regeneration of the mangroves. The sea level rise is also affecting the availability of sediment, directly impeding the establishment of new groves. In 2010 it was observed that, with the sea levels rising ominously, the New Moore Island/ South Talpatti Island disappeared. Scientists have predicted that the other low-lying islands could also vanish as sea levels continue to rise.

C. Cyclones

It has been observed that there has been an increase in the intensity of cyclonic storms hitting the Sundarbans between 1951 and 2010. Recent research suggests that such increase in intensity may be attributed to an increase in sea surface temperature. During the occurrence of Cyclone **Aila** in 2009, a 400 km stretch of embankment was breached and the waves crossed over the height of the embankment and entered the flood plains. The sea water got logged and more than 2 million people were marooned for several days affecting not only agriculture but also drinking water supply. Most of the thatched houses were damaged. Additionally, the farmland became nonproductive due to salt water incursion in the soil. Frequent damage to property and crop failure due to cyclones and thunderstorms has rendered a large population to be poverty ridden. As a result, high levels of immigration to cities are reported from here. The recent cyclones, **Bulbul**, and obviously, the latest one, **Amphan** (in May, 2020) has left a completely devastating impact on Sundarban's life.

Source: lifegate.com

Figure 7: Effect of the Super-cyclone Amphan in Sundarbans

D. Rise in Salinity; Impact on Agriculture

Studies suggest that in the last two decades, the run off in the eastern rivers of Sundarbans has decreased resulting in ever increasing salinity and sea water-sulfate concentrations. The decrease in fresh water run-off has affected mangrove growth. Further, agriculture is being affected because of the high levels of salinity of the soils due to high tides, cyclones and storm surges, and problems of water stagnation, sometimes even beyond monsoon seasons.

E. Change in Agricultural Patterns

According to the West Bengal Government, in 2009, the agricultural area had shrunk between 2002 and 2009 from 2,149.615 square kilometers to 1,691.246 square kilometers. The area suffers from a low intensity of cropping because mono cropping of rice is practiced seasonally, and horticultural crops are rarely grown. Further, only 12% of the cropped area in the Sundarbans is irrigated through rainfed ponds, tanks and canals; majority of the agricultural land is rainfed. It has been observed that rainfall has become erratic and its intensity has increased causing further damage to the agricultural yield. With continuous increase in population, agriculture production in the region is not able to meet demand. Historically, the main economic activity of rain-fed paddy agriculture was made possible by the construction of earthen embankments to keep brackish tidal water at bay and by cultivating salt-tolerant paddy varieties such as *Matla* and *Hamilton*. Such varieties could be cultivated on raised sections of the islands even without embankments. However, while searching for the six locally recalled salt-tolerant farmers' paddy varieties; scientists have discovered the availability of only two varieties; while the other salt resistant rice varieties are believed to have been lost under the onslaught of India's "green revolution".

F. Deforestation

Between 1777 and 1971, continuous deforestation and land reclamation activities have been carried out in the Sundarbans region. It has been observed that 5% of forest cover was lost between 1989 and 2009. This deforestation has increased man-animal conflict, local extirpation of several species and added to the biological loss of the region. Further, clearing of forests have not facilitated self-sustaining agriculture on the flood plain, as it tends to be submerged under saline water during high tides.

G. Pollution

Due to heavy siltation and disposal of solid waste from adjacent cities, the rivers in the Sundarbans do not receive fresh water from the upstream Ganges and are primarily tidal fed. The Sundarbans receives an additional supply of fresh water during monsoon which lasts for a period of May to October. A pronounced ecological change has evolved in this delta due to huge discharges of untreated domestic and industrial effluents carried by tributary rivers as well as the disposal of contaminated mud from harbour dredging and from the rapid emergence of the Haldia Port Complex, a major oil disembarkment terminal in eastern India. The Sundarbans delta has become susceptible to chemical pollutants such as heavy metals which may have changed the estuary's geochemistry and affected the local coastal environment. Due to a diversity of inputs such as agricultural runoffs, wastewater and sewage discharges, and agricultural wastes, maximum concentrations of organochlorine pesticide residues were recorded at sites located along the main stream of the Hugli (Ganges) estuary. This has a disastrous impact on the biodiversity harboured by mangroves.

Conservation Efforts

Conservation of the Sundarban mangrove is supposed to have started with its declaration as a reserve forest, under the Indian Forest Act in 1878, after Schlich (1875) raised concern over its conservation (Presler, 1991). The hunting of tigers was banned completely in 1970, after the IUCN listed the Bengal tiger (*Panthera tigris tigris*) as an endangered species. Later, under Project Tiger, the Government of India established a Tiger Reserve in the Sundarban covering 2,585 km2 in 1973 (Fig. 4a). Another 241 km2 area was demarcated as a subsidiary wilderness area. The core area of 1,330 km2 was later designated as a National Park. Another wildlife sanctuary was established in 1976 on Haliday Island (595 ha) to protect the spotted deer (*Axis axis*), wild boar (*Sus scrofa*) and rhesus macaque (*Macaca mullata*), which are dominant animals in a forest type consisting mainly of *Ceriops decandra*. In 1987, the Sundarban National Park in India, and in 1997, parts of the Sundarban in Bangladesh, were inscribed on the World Heritage list (IUCN, 1997). Within the Biosphere Reserve several distinct zones have been recognised: A Core Zone comprised of the National Park and the Tiger Reserve, a Manipulation Zone (2,400 km2 of mangrove forests), a Restoration Zone that covers 240 km2 of degraded forest and saline mud flats, and a Development Zone which includes mostly the reclaimed areas. Only the Core Zone is under strict conservation measures. Income generating activities such as the collection of seeds of black tiger prawn (*Penaeus monodon*), the culturing of oysters and crabs, mushroom cultivation and bee-keeping for honey production are allowed in the Manipulation Zone. Efforts are being made, however, to also rehabilitate certain degraded areas through afforestation. Among faunal species, the estuarine crocodile and the Olive Ridley turtle are receiving some attention by way of captive breeding. In India however, the strategy for Sundarbans conservation involves the setting aside of areas where the entire life-cycle needs of a community can be met and the ecological needs of wildlife can be linked into the overall management of the system for the Sundarbans. Furthermore, in so doing, the ecological processes upon which wildlife depend become integral values in the management matrix.

Future

Considering the humongous damage and widespread devastation caused by Super Cyclone Amphan, following steps should be taken to prevent further degradation of Sundarbans Biosphere Reserve, the World Heritage Site and the associated ecologically fragile landscapes of the region which makes life and livelihoods of the stakeholders vulnerable. The reasons for increased vulnerability of Sundarbans have been studied by (Mitra, *et al.*, 2011) which includes influx of coastal water and inundation of forested areas due to shallow bathymetry extending tens of kms offshore, bay convergence, tidal influence, densely populated low-lying islands with poor connectivity with mainland, favorable cyclone tracks making landfall perpendicular to coastline (as was observed during Aila and Amphan) as well as crucial interlinking of innumerable inlets and river systems contributing to easy entry of storm-water, saline water ingression due to easy backflow of sea water through the streams. The intertidal mudflat in front of the embankment should be planted to grow wild rice variety, *Porterasiacoarctata* (Syn = *Oryzacoarctata*). This species is a tetraploid wild rice variety growing abundantly across the coastal region of India and especially can be encouraged in Sundarbans. The salt tolerance property of this mangrove associate has been dealt with by a number of workers earlier (Sengupta and Majumdar, 2010). Further, plantation of this rice variety would ensure stabilization of the temporary embankments constructed by villagers every year during the Pre-Monsoon periods.

Similarly, geo-jute must be applied to cover the boulders supporting the embankments. Plantation of mangroves should be carried out on the geo-jute substratum with an allowance of sufficient root penetration in the soil. 40% of the total biomass should be allocated to belowground structures to ensure stability of the mangroves. So, during mass scale mangrove afforestation programme in the coming months as is announced by Government of West Bengal, species of mangroves are to be chosen very carefully to ensure maximum stability of the soil, prevention of soil erosion, water recharging as well as minimum run-off during monsoonal rains. Moreover, focus should be on the plantation of fast-growing species that could mitigate the vulnerabilities of the region during the Post-Amphan periods. This would further ensure sustainable afforestation with minimal damage during similar such natural disasters in future. Securing the future of the Sundarbans, its biodiversity and people requires a long-term vision that can integrate climate adaptation and conservation strategies along with shorter term interventions such as ensuring sustainable livelihoods, access to clean and sustainable energy and effective human wildlife conflict management.

References

Alongi, D. M. 2002. Present state and future of the world's mangrove forests. Environmental Conservation, 29: 331–349.

Bhavan, Aranya. "Training Workshop on 'Fundamentals of Global Warming & Climate Change with Special Reference to Sundarbans Ecosystem.'" Sundarbans Biosphere Reserve, Sundarbans Biosphere Reserve, Directorate of Forests, West Bengal, 6 Dec. 2010.

Chakrabarti, Ranjan. "Local People and the Global Tiger: An Environmental History of the Sundarbans." Global Environment, vol. 2, no. 3, 2009, pp. 72–95.

Danda, Anurag. Sundarbans: Future Imperfect – Climate Adaptation Report. WWF-India, 2010, pp. 1–28, Sundarbans: Future Imperfect – Climate Adaptation Report.

Das Gautam Kumar, Biodiversity of Sundarbans; Environment and Biodiversity, lambert Academic Publishing, 2010.

Dasgupta N. Nandy P, Sengupta C and Das S, "Salinity Impact on the Precarious Mangroves: A Biochemical Study on Some Taxa from Indian Sundarbans," American Journal of PlantPhysiology, vol. 7, pp. 53-69, March 12, 2012.

Jalais, Annu. Forest of Tigers People, Politics and Environment in the Sundarbans. Taylor and Francis, 2014.

Mitra Abhijit et al, Amphan Supercyclone: A death knell for Indian Sundarbans, eJournal of Applied Forest Ecology (eJAFE), Vol.8, No.1 (2020) 41-48.

Mahadevia K., Bikas M., Climate Change: Impact on the Sundarbans; International Scientific Journal of Environmental Science.

5

Environmental and Social Impact of Super Cyclone Amphan

Dr. Savita Mishra
Principal, Vidyasagar College of Education, Phansidewa
Darjeeling, West Bengal, India

Abstract

Cyclone Amphan is a tropical cyclone formed over the Bay of Bengal that has intensified and likely to turn into a "super cyclonic storm (maximum wind speed is 224 kmph)". It has been named by Thailand. This is the first super cyclone to form in the Bay of Bengal after the 1999 super cyclone that hit Odisha and claimed more than 10,000 lives. It is the third super cyclone to occur in the North Indian Ocean region after 1999 which comprises of the Bay of Bengal, the Arabian Sea and the northern part of the Indian Ocean. The other two super cyclones were Cyclone Kyarr in 2019 and Cyclone Gonu in 2007. Super Cyclonic Storm Amphan is a powerful and deadly tropical cyclone that caused widespread damage in Eastern India, specifically West Bengal, and also Bangladesh in May 2020. Cyclone makes landfall, slamming into India's coast during a pandemic, knocking down huge trees, bringing ropes of rain and sending millions of poor villagers rushing into evacuation shelters. As a result of climate change, ocean temperatures were raised. Thousands of trees uprooted destroying environment and livelihood. The heavy rainfall brought by the cyclone uprooted over 5,000 trees as well as several lampposts and traffic signals across the state. This chapter deals with the damages done by cyclone Amphan, environmental, economic, social, educational, psychological and health impact and towards improvement with suggestive measures.

Keywords: Cyclone, Amphan, Environmental Impact, Social Impact.

Introduction

Cyclones are formed over slightly warm ocean waters. The temperature of the top layer of the sea, up to a depth of about 60 meters, need to be at least 28°C to support the formation of a cyclone. This explains why the April-May and October-December periods are conducive for cyclones. Then, the low level of air above the waters needs to have an 'anticlockwise' rotation (in the northern hemisphere; clockwise in the southern hemisphere).

Grading of Cyclones

Tropical cyclones in the Bay of Bengal are graded according to maximum wind speeds at their centre. At the lower end are depressions that generate wind speeds of 30 to 60 km per hour, followed by:

1) **cyclonic storms** (61 to 88 kmph)
2) **severe cyclonic storms** (89 to 117 kmph)
3) **very severe cyclonic storms** (118 to 166 kmph)
4) **extremely severe cyclonic storms** (167 to 221 kmph) and
5) **super cyclones** (222 kmph or higher)

Cyclone Amphan

Cyclone Amphan is a tropical cyclone formed over the Bay of Bengal that has intensified and likely to turn into a "super cyclonic storm (maximum wind speed is 224 kmph)". It has been named by Thailand. Amphan is the equivalent of a Category 5 hurricane on the Saffir-Simpson Hurricane Wind Scale but when it made landfall in West Bengal, Amphan tone down into a category 4 Extremely Severe Cyclonic (ESC) storm with a wind speed of 165-175 kmph.

Amphan in Brief

Formed: 16 May 2020

Dissipated: 21 May 2020

Highest winds: 3-minute sustained: 240 km/h (150 mph) 1-minute sustained: 260 km/h (160 mph)

Lowest pressure: 920 hPa (mbar); 27.17 in Hg

Fatalities: 128 total

Damage: >$13.6 billion (2020 USD) (Costliest on record in the North Indian Ocean)

Areas Affected: India (West Bengal, Odisha, Andaman Islands), Bangladesh, Sri Lanka, Bhutan and Part of the 2020 North Indian Ocean Cyclone Season.

Figure 1

Amphan: What Makes it a Nightmare?

This is the first super cyclone to form in the Bay of Bengal after the 1999 super cyclone that hit Odisha and claimed more than 10,000 lives. It is the third super cyclone to occur in the North Indian Ocean region after 1999 which comprises of the Bay of Bengal, the Arabian Sea and the northern part of the Indian Ocean. The other two super cyclones were Cyclone Kyarr in 2019 and Cyclone Gonu in 2007. Super Cyclonic Storm Amphan is a powerful and deadly tropical cyclone that caused widespread damage in Eastern India, specifically West Bengal, and also Bangladesh in May 2020. It is the strongest tropical cyclone to strike the Ganga Delta since Sidr of the 2007 season and the first super cyclonic storm to have formed in the Bay of Bengal since the 1999 Odisha cyclone. It is also the 3rd super cyclone that hit West Bengal since 1582, after 1737 and 1833, as well as being the strongest. Causing overUS$13 billion of damage, Amphan is also the costliest cyclone ever recorded in the North Indian Ocean, surpassing the record held by Cyclone Nargis of 2008.

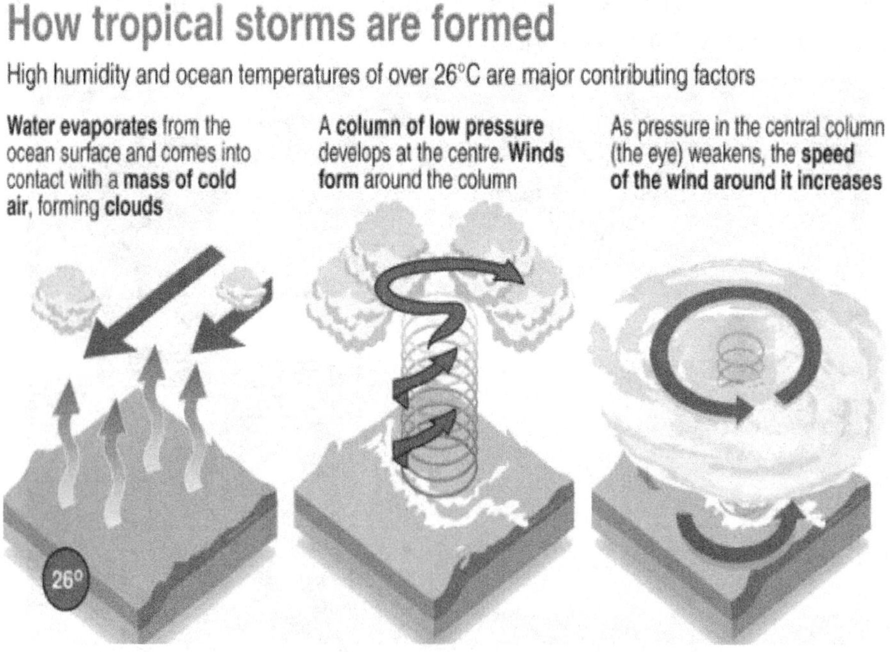

Figure 2: Formation of Tropical Storms

The first tropical cyclone of the 2020 North Indian Ocean Cyclone Season, Amphan originated from a low-pressure area persisting a couple hundred miles (300 km) east of Colombo, Sri Lanka, on 13 May 2020. Tracking northeastward, the disturbance organized over exceptionally warm sea surface temperatures; the Joint Typhoon Warning Centre (JTWC) upgraded the system to a tropical depression on 15 May while the Indian Meteorological Department (IMD) followed suit the following day. On 17 May, Amphan underwent rapid intensification and became an extremely severe cyclonic storm within 12 hours. On 18 May, at approximately 12:00 UTC, Amphan reached its peak intensity with 3-minute sustained wind speeds of 240 km/h (150 mph), 1-minute sustained wind speeds of 260 km/h (160 mph), and a minimum central barometric pressure of 920 mbar (27.17 in Hg). The storm began an eyewall replacement cycle shortly after it reached its peak intensity, but the continued effects of dry air and wind shear disrupted this process and caused Amphan to gradually weaken as it paralleled the eastern coastline of India. On 20 May, between 10:00 and 11:00 UTC, the cyclone made landfall in West Bengal. At the time, the JTWC estimated Amphan's 1-minute sustained winds to be 155 km/h (100 mph). Amphan rapidly weakened once inland and dissipated shortly thereafter. Coastal areas in West Bengal comprising East Midnapur, North 24 Parganas, South 24 Parganas, Kolkata, Hoogly and Howrah as well as Odisha were affected by the cyclone. It also caused significant destruction in Bangladesh.

Trails of Destruction: West Bengal

Amphan made landfall near Bakhali in West Bengal at 2:30 p.m. IST on 20 May, buffeting the region with strong winds and heavy rains. Although the extent of fatalities was less than initially feared, the cyclone's effects were nonetheless widespread and deadly. West Bengal, the epicenter of the cyclone's landfall, saw the most widespread damage from Amphan. The storm was considered the strongest to hit the region in over a decade. At least 86 people died in West Bengal; most of the fatalities were due to electrocution or the collapse of homes. The state government estimated that the storm caused at least 1 trillion (US$13.2 billion) in damage and directly affected 70 percent of the state's population. Chief Minister Banerjee described the storm's effects there as worse than that of COVID-19. An estimated storm surge of 5 m (16 ft) inundated a wide swath of coastal communities and communications were severed. The greatest inundations were expected in the Sundarbans, where flooding could extend 15 km (9.3 mi) inland. Bridges linking islands to the Indian mainland were swept away. The cyclone produced sustained winds of 112 km/h (70 mph) and gusts to 190 km/h (120 mph), which were recorded by the Alipore observatory, Kolkata, West Bengal, damaging homes and uprooting trees and electric poles. Wind speed along coastal areas were measured up to

150–160 km/h (93–99 mph). This overturned vehicles and snapped approximately 10,000 trees. The Calcutta Municipal Corporation stated that Amphan toppled over 4,000 electric poles, leaving much of the city without power for over 14 hours. At least 19 people were killed in Kolkata. The storm also triggered widespread flooding around the city. 236 mm of rain was recorded in Kolkata.

Odisha

Neighbouring Odisha saw significant effects, with wind speed reaching 106 km/h (66 mph) and rainfall up to 300 mm (12 in) in Paradip. In Bhadrak, rainfall reached 384.6 mm (15.14 in). Approximately 1,167 km (725 mi) of power lines of varying voltages, 126,540 transformers, and 448 electrical substations were affected, leaving 3.4 million without power. Damage to the power grid reached 3.2 billion (US$42 million). Four people died in Odisha, two from collapsed objects, one due to drowning, and one from head trauma. Across the ten affected districts in Odisha, 4.4 million people were impacted in some way by the cyclone. At least 500 homes were destroyed and a further 15,000 were damaged. Nearly 4,000 livestock, primarily poultry, died. The Cyclone was strongest at its northeast section.

Southern India

Rains and strong winds from Amphan swept across many districts in Kerala beginning on May 16. Thunderstorms associated with Amphan caused severe coastal erosion in the Valiyathura suburb of Thiruvananthapuram, damaging roads and destroying homes and threatening to displace over a hundred families from their homes. Tamil Nadu faced some impact from the cyclone. Heavy winds damaged at least 100 boats anchored in the Ramanathapuram district. Strong winds inflicted severe damage in Kottayam district, especially in Vaikom taluk, where homes and temples were impacted and trees and electric poles were downed. In Sooradapeta, near Kakinada in Andhra Pradesh, rough seas destroyed 35 homes and damaged several others.

Bangladesh

Over a million people were affected by Amphan across nine districts in the divisions of Khulna and Barisal. At least twenty people died in storm-related incidents. Nearly 220,000 homes were damaged, of which 55,667 were destroyed, rendering an estimated 500,000 people homeless according to the United Nations Office for the Coordination of Humanitarian Affairs. Three hundred shelters in Cox's Bazar were damaged by flooding and landslides. Across 26 districts, approximately 1,100 km (680 mi) of roads and over 200 bridges were damaged.

Sri Lanka

The cyclone produced heavy rainfall and strong winds in Sri Lanka while intensifying east of the island, affecting some 2,000 people and triggering floods and landslides. Two people were killed as a result of these rains in Ratnapura District, with one killed by a landslide and another by a fallen tree. Landslide-related injuries hospitalized other residents in the area. Two people were killed in Kegalle, where 214 mm (8.4 in) of rain fell in 24 hours. Flash floods in Kottampitiya and Pelmadulla prompted the evacuation of 60 people from homes susceptible to a possible landslide. Over 500 homes were damaged by Amphan, of which 145 were in Polonnaruwa.

Bhutan

The remnants of Amphan produced several days of unsettled weather across Bhutan, with the Tsirang District in particular seeing three days of heavy rain. Landslides and rockfalls across the nation blocked roadways and damaged homes. Flash flooding in Threna damaged crops and prompted the evacuation of five households. Seventeen yaks and horses died across Lingzhi, Naro, and Soe gewogs. Stemming from damage to fibre-optic cables in Kolkata, Bhutan Telecom saw a total loss of service for 17 hours on 20–21 May and TashiCell saw a 60 percent outage. Power outages occurred in Drepong, Gongdu, Iurmey, Kengkhar and Silambi gewogs.

Here is a List of Damages Expected, as Released by the IMD

Extensive damage to all types of kutcha houses, some damage to old badly managed pucca structures, potential threat from flying objects, extensive uprooting of communication and power poles, disruption of rail/road link at several places, extensive damage to standing crops, plantations, orchards, blowing down of palm and coconut trees, uprooting of large bushy trees and large boats and ships may get torn from their moorings.

Precautionary Measures (India)

Prime Minister Narendra Modi held a meeting with Minister of Home Affairs Amit Shah, Chief Minister of West Bengal Mamata Banerjee, Chief Minister of Odisha Naveen Patnaik and other officials on 18 May to review preparations and evacuation plans. Teams from the Odisha Disaster Rapid Action Force and National Disaster Response Force (NDRF) were pre-positioned across districts in Odisha and West Bengal on 17 May to assist in preparations for Amphan and render aid where necessary, with additional units placed on standby; these units could be readily airlifted to affected areas onboard Indian Air Force transport aircraft. The National Disaster Management Authority advised that these crews and other first responders would also need personal protective equipment's and N95 masks due to the ongoing pandemic. A diving team from the Indian Navy was sent to Kolkata to aid relief efforts, with ships from the navy placed on standby for relief operations. The government of Odisha directed the magistrates of four districts on 15 May to establish shelter homes for possible evacuees. Odisha Chief Secretary Asit Kumar Tripathy initially identified 403 possible cyclone shelters in areas potentially impacted by Amphan, though 105 served as temporary medical centres for quarantines associated with the concurrent COVID-19 Pandemic. Shelters could only be filled to one-third capacity to maintain social distancing guidelines due to the pandemic. Social distancing restrictions in West Bengal reduced evacuation capacity in shelters from 500,000 people to 200,000 people. The Kolkata Municipal Corporation located schools and community centres for possible use as temporary shelters to augment evacuation capacity. At least 1,704 shelters were ultimately established in Odisha and more than 2,000 were used in West Bengal, including schools and public buildings.

Precautionary Measures (Bangladesh)

The Bangladesh Cyclone Preparedness Programme and National Disaster Response Coordination Group convened on 18 May to outline preparations for Amphan. Several non-governmental agencies coordinated with the Cyclone Preparedness Programme to support Rohingya refugees at camps in Cox's Bazar during Amphan. All 32 Rohingya camps in Cox's Bazar were staffed with relief volunteers. Health services in Bangladesh during the storm were reinforced by 1,933 medical teams distributed around Bangladesh. Fifteen thousand volunteers and 284 medical teams were prepared to render aid around Chittagong. The Bangladesh Cyclone Preparedness Programme and National Disaster Response Coordination Group convened on 18 May to outline preparations for Amphan. Several non-governmental agencies coordinated with the Cyclone Preparedness Programme to support Rohingya refugees at camps in Cox's Bazar during Amphan. All 32 Rohingya camps in Cox's Bazar were staffed with relief volunteers. Health services in Bangladesh during the storm were reinforced by 1,933 medical teams distributed around Bangladesh. Fifteen thousand volunteers and 284 medical teams were prepared to render aid around Chittagong. More than 12,000 shelters were opened across Bangladesh, supplied with food and emergency cash from the Bangladeshi Ministry of Disaster Management and Relief yielding capacity for 5.19 million evacuees. There were 5,767 more shelters in Bangladesh for Amphan than for Cyclone Bulbul, which struck the country in November 2019, due in part to social distancing restrictions. Approximately 4.2 million people were evacuated in coastal India and Bangladesh, with roughly 2 million from India and 2.2 million from Bangladesh. Most of the evacuations in India occurred in West Bengal. Over a million people were expected to evacuate from areas near the Bangladesh-India border. Around 4,000 personnel from the SDRF oversaw evacuations in India. Evacuations began on 17 May in Jagatsinghpur, beginning with the elderly and pregnant living in thatched homes.

Precautionary Measures (Sri Lanka, Thailand, Bhutan)

Flood and landslide warnings were issued by the Disaster Management Centre of Sri Lanka on 19 May. Boats, helicopters, and crews were stationed by the Sri Lanka Air Force and Sri Lanka Navy throughout the island to bolster emergency response. Coastal provinces of Thailand along the Andaman Sea were warned by the Thai Meteorological Department of the threat of heavy rain on 19 May. Storm alerts issued by the Thai Department of Disaster Prevention and Mitigation encompassed 62 provinces, including Bangkok, while warnings for flash floods, high waves, and landslides were issued for 14 provinces in Southern Thailand. The National Centre for Hydrology and Meteorology of Bhutan warned of the possibility of flash flooding and landslides in the country. The Bhutanese Department of Disaster Management advised against travelling, particularly in southern Bhutan.

Aftermath

Almost 1 lakh houses have been destroyed and over 1 crore people have been affected by Amphan, according to the Bengal government. Electricity supply was disrupted in certain coastal areas after electric poles were uprooted by strong winds. In other districts of south Bengal, there were reports of old houses collapsing and trees being uprooted. On 22 May, Prime Minister Narendra Modi conducted an aerial survey over Kolkata, along with Chief Minister Mamata Banerjee. Modi announced a 10 billion (US$132 million) immediate relief package for West Bengal and 5 billion (US$66.2 million) in relief for Odisha. In advance, Modi announced that 200,000 (US$2,650) would be provided to the next of kin of people who died during the storm, and 50,000 (US$660) would be given to each injured person. West Bengal CM Banerjee stated that it would take three to four days to assess the damage. Approximately 1,000 ground teams worked to restore infrastructure and services in West Bengal after Amphan, though only 25–30 percent of workers were staffed due to the COVID-19 pandemic. The resulting slow restoration of power sparked protests across West Bengal aimed primarily at electricity company CESC. The Home Department of West Bengal requested additional crews from railway and port interest, while five brigades from the Indian Army were deployed in Kolkata and the 24 Parganas districts to support recovery efforts. The government of Odisha sent 500 members of its disaster rapid action force and fire service to West Bengal.

Environmental Impact

Cyclone makes landfall, slamming into India's coast during a pandemic, knocking down huge trees, bringing ropes of rain and sending millions of poor villagers rushing into evacuation shelters. As a result of climate change, ocean temperatures were raised. Thousands of trees uprooted destroying environment and livelihood. The heavy rainfall brought by the cyclone uprooted over 5,000 trees as well as several lampposts and traffic signals across the state. At least 400 percent raised in the pollution level against the permissible limit of World Health Organization (WHO) as most of the green cover of the city and districts have been lost in the cyclone and increased pulmonary diseases. River embankments have been breached in South and North 24 Parganas districts in West Bengal, leaving several houses inundated as water through the breaches slipped into remote settlements, mostly in the Sundarbans. While most districts of West Bengal are impacted, the Sundarbans region is the worst hit because it is an ecologically fragile zone.

1) **Bengal Recorded Agri-Loss Due to Cyclone Amphan:** Cyclone Amphan has ravaged the farming sector in at least 14 of the 23 districts in West Bengal and the loss incurred may create an all-time record according to, state Agriculture Minister Ashish Banerjee. According to initial estimates, Hooghly and Birbhum districts have incurred a loss of Rs 600 crore and Rs 462 crore respectively.

2) **Amphan Damaged World's Largest Banyan Tree in Howrah Botanical Garden:** The Amphan cyclone which hit West Bengal a day before the International Day for Biological Diversity on Friday has severely damaged the world's largest banyan tree in the Howrah botanical garden. This cyclone has also uprooted one of the two 'kalpvriksha', having botanical name as olea cuspidata and also known as wild olive, brown olive or Indian olive, in the Acharya Jagadish Chandra Bose Indian Botanic Garden in Howrah, Allahabad-based scientist Shiv Kumar of Botanical Survey of India on Friday. Experts say that the Sundarbans – the world's largest mangrove forest, on a vast delta that supports several hundred animal species —could act as a buffer against the cyclone when it strikes, slowing wind speeds and protecting villages from the worst effects. But around 70 percent of the 4,000-square-mile forest, which spreads across the border between India and Bangladesh, is just a few feet above sea level. Heavy rains and flooding could have deadly consequences for rare animals there, pushing them out of their habitats and into areas populated by humans. Cyclone Amphan poses a particular

risk for the region's few hundred Bengal tigers, one of the largest remaining populations of wild tigers in the world. Habitat loss, hunting and the illegal trade of animal parts have devastated the global population of tigers— now estimated at fewer than 4,000, compared with about 100,000 a century ago.

Economic Impact

The European Union stated that it would initially provide €500,000 (US$545,000) for those affected by the storm in India. The Ministry of Disaster Management and Relief in Bangladesh approved a 2.5-3 billion (US$29–35 million) budget to repair embankments damaged by Amphan. Another 1.5 billion (US$18 million) was distributed to each district heavily impacted by the storm, along with 500 bundles of corrugated tin sheets. International development organization BRAC disbursed 30 million (US$350,000) to low-income families in 10 upazilas, with 5,000 (US$60) per family. The organization also distributed gloves, masks, and sanitizers to Bagerhat, Khulna, and Satkhira districts.

Social Impact

From a social perspective, it is evident that the poorest communities in the Sundarbans and rural Bengal have borne the brunt of Cyclone Amphan-they have lost their homes, and their fields are inundated with saltwater and are rendered useless in the near future. This is mirrored by natural disasters around the world. Any future climate adaptation strategy needs to account for the material needs of the most vulnerable communities, prioritizing the building of a social safety net that will enable them to resume their livelihoods and continue to live with dignity. Moreover, among the various stakeholders that build and carry community resilience, grassroots organizations and volunteer networks understand the local community and show up in solidarity at the hour of need, as disasters around the world have shown time and again. In Bengal, organic volunteer networks like the Quarantined Student Youth Network and the Bengal Relief Collective, and grassroots groups like Mukti are among several groups that have rapidly mobilized resources and deployed volunteers on the ground to provide aid to affected communities. The Indian government, as well as the West Bengal government had taken steps to ensure the safety of both life and property in the states. About three lakhs people had been evacuated from the region's most at risk. Immediate trauma, injuries, loss of life and property were the main focus of medical relief measures. However, it's important to remember that health systems and facilities, which are already spread thin due to the COVID-19 pandemic, will also have to deal with the long-term health impacts of a severe storm like this. Officials in Bhadrak, a city in the Indian state of Odisha, have evacuated 218 Pregnant women from the area's coastal villages to health centres, 60 of the evacuated women gave birth on May 19 and 20 under special medical care, as officials braced for the cyclone. Thousands of Indian migrant workers attempting to return home to the country's Odisha state remain stranded after the cyclone forced the cancelation of special trains to the region.

Social Impact (Bangladesh)

Women and girls lived in the Cox's Bazar refugee camp in Bangladesh were vulnerable to an increase in gender-based violence, as the cyclone and coronavirus threaten their security. The sprawling camps of Cox's Bazar are home to nearly one million stateless Rohingya refugees. Life could not be more precarious for the nearly half a million women and girls who are already trapped in their homes as well as the children who make up over half of the camp's residents. Many of them have experienced gender-based violence, often at the hands of their partners, and now they risk having their homes uprooted to be forced into more condensed spaces. An "unimaginable" strain put on relationships and families in the camp "and there was a real risk that gender-based violence increased due to Amphan cyclone."

Educational Impact

Schools and Colleges were already shut down due to COVID-19, Infrastructures were also destroyed and damaged due to Amphan. No facility of online education due to lack of proper infrastructure, electricity, Internet etc. No contact with teachers and students due to two disasters at a time-COVID-19 and Amphan cyclone. Lack of proper communication was due to damage of electric poles. All educational and research activities were stopped, all conferences were postponed due to Amphan cyclone.

Health Impact

The following are the five major health impacts due to Amphan cyclonic in worst-hit states.

1) **Long-Term Trauma:** World Health Organization (WHO) categorized as "indirect impact" due to loss of life, shelter and livelihoods for short- and long-term mental health effect. This can potentially delay rehabilitation of affected areas and return to normalcy, especially given that the novel corona virus is still a matter of grave concern.
2) **Food Shortage:** Lack of or disrupted food supplies in affected areas could lead to long-term food shortages and malnutrition, especially among newborns, infants and the older populations. This can raise mortality rates in the coming months as a result of the indirect impact.
3) **Water Contamination:** Contamination of water with faeces, silt and debris were likely to affect all cyclone-hit areas, leading to a rise in cases of diarrhea, dysentery and other water-borne diseases like cholera and typhoid fever.
4) **Communicable Diseases:** Communicable and infectious diseases are most likely to rise after a major cyclonic storm or typhoon. Water-logging after a cyclone might also lead to outbreaks of vector-borne diseases like malaria, dengue, chikungunya, etc. in the affected areas. Number of skin infection cases also goes up.
5) **Non-Communicable Diseases:** Non-communicable diseases took a major hit after a Amphan cyclonic due to disruptions in public health services. This leads to exacerbated illness and even death due to ailments like cancer, hypertension, diabetes, cardiovascular disease, kidney disease, etc. A cyclonic storm can, therefore, increase the disease burden of any number of ailments affecting the people residing in the worst-hit areas in the long run.

Psychological Impact

Due to the heavy damages, loss of shelters, lack of food, medicines etc. people were suffered from mental health problem, stress, anxiety, depression, fear etc.

Towards Improvement

Precautionary Measures were already taken by all government and authorities to save people from Amphan cyclone. Early Warning and Dissemination of Local People were given before the arrival of cyclone Amphan, Shifting of people to safe places, QRT (QUICK RESPONSE TEAM) at local level were active and started working day and night to save people, started monetary help, arranged Food, Shelter, Sanitation, Medicines etc. for people affected by cyclone Amphan, given update information of happening around them, resettlement: Safe returning, Back to normal life, Fast relief packages after proper assessment to right and needy person, started reconstruction of Public and Private Property, Environmental care, Social safety, Providing health facilities and Psychological safety and proper counseling.

References

Climate Adaptation Planning, by Avantika Goswami, May 26, 2020.

CNN's Salman Saeed.

CNN's Manveena Suri in New Delhi.

Dibyangshu Sarkar/Agence France-Presse — Getty Images.

Enviro & Biodiversity, Mains Paper 1: Geographical Features & Their Location, Postdate May 19, 2020, Postdate The Hindu.

French Press Agency-Afp, West Bengal/Kolkata Pacific, May 19, 2020 4:26 Pm Gmt+3.

Inputs from PTI.

Suzauddin Rubel/Agence France-Presse — Getty Images.

Web Desk. May 18, 2020 11:22 IST.

6

Microclimatic Effect on Tea Cultivation: Rain Fall and Yield

Dr. Ratula Mukherjee

Assistant Professor, Department of Botany,
Ananda Chandra Training College, Jalpaiguri, West Bengal, India

Abstract

Since the 18 century, Indian economy has been experiencing a high profit export and earnings from the cash crop especially the Tea. Initially the states of Assam and Northern parts of Bengal during the British rule were the geographic areas of Tea production. After the independence and the initiative taken by the Tea Board and the Ministry of Commerce, Government of India, the geographic areas of tea production have been extended to the Himachal Pradesh and Southern India. The tea production needs specific geographic and environmental conditions as well as some cultural and agricultural practices. Many more researches have been conducted on the different aspects of geographical and environmental requirements especially the significance of microclimate. And it is continued. The present study tried to find out the correlation between the rain fall and the yield of Tea crops in a specific area of Northern West Bengal, i.e. Dooars region (delimitation - Alipurduar). The study has been conducted by acquiring the rain fall data and yield of a tea garden lying amidst of Dooars, as representative one. Those were the secondary data and the Excel data analysis tool was applied as statistical method. The rain fall data utilized in the study were average as the garden was there in the central position of Dooars.

Keywords: Tea as Cash Crop, Rain Fall Per Anum, Yield (Production) Per Anum, Correlation of Two Variables.

Introduction

Since the centuries, a warm, aromatic brew and just sipping it from the bowl or cup, have not only been touching the taste buds of the people, the heart and soul in addition. The decades after decades have gone, and the crop became the myth. A complete untold story is there which is seldom unveiled. The story tells about the Tea. So many researchers work and studies are there to unveil the consequences of the environmental requirements that affect the tea cultivation.

Tea needs huge rainfall (about 250 cm or more per annum) beyond which the yield suffers. But more precisely to say, Tea cultivation needs a regular pattern of rainfall. Here in Dooars region, the rain fall starts from April-May, the starting of summer which continues periodically throughout the year along with the monsoon even in the December (called winter rain). While the monsoon rain has a significant role for tea yield and deposition of soil water, the winter rain has tremendous effect on the deposition of sub soil water, survival of pruned tea bushes. Even a minimum rain (0.5-1.5 inch) in February and March has a significant role in the yield and overall production.

The soil water availability is crucial for the administering of fertilizers and manures. Reducing volume of rainfall results the water stress condition, especially in the subsoil up to the level of 36-40 inches or even more. Generally, the roots of clonal varieties of tea plants (e.g. TV 21, TV 23) go down within this level. Therefore stress of water in this level causes severe harm to the plants, generally the young tea plants. These plant varieties are found suffering from mineral nutrients due to the inability to absorb and translocation of the same through the water medium.

The rainfall has a significant effect on the pest infestation. The major pests in this region are Looper (*Buzura suppresuria*) and *Helopeltis* sp. along with Red Slug Cater pillar. Reduce rainfall or irregular pattern causes severe pest attack resulting huge crop loss. Severe infestation demands ultimately high utilisation of chemical pesticides. But, in the water stress condition, systemic pesticides cannot be utilised and thus the recurring use of contact pesticides would be practiced that lead to an expensive practise.

Here the study has tried to find the effect of the rain fall on the tea cultivation in terms of yield in the region of Dooars of North Bengal. It is revealed that the environment has been changing rapidly which affects the rainfall pattern throughout the region over last few years. The study has depicted so. There are so many reason of such changing pattern of the environment. Urbanization, deforestation, degradation of the rivers' basin and so many reasons are there which affect the biosphere in micro level adversely.

Objectives of the Study

The main objectives of the study were to show:

1) The rain fall pattern of the specific area as one variable (independent), the yield of the specific garden as second variable (dependent) and find out the correlation between the two variables, whether weak, moderate, strong significance or not.
2) More over the study analyzed the rain fall pattern and tried to find out whether any irregular pattern was there.

Hypothesis

H0 (Null): There is no significance/ relation between the rain fall variation and yield of the tea.

H1(Alternate): There is significant relationship between the two variables.

Methodology

- Collection of data of two variables from the source (secondary data) [data covered from the year 2010 to 2015].
- Present the data in tabular form.
- Use Excel data analysis tool to find out the correlation.

Rainfall and Yield of Crop

Within the year 2010 to 2015 (the data time), the rainfall variation pattern in the Dooars (Hasimara-Alipurduar) region (study area) has shown irregularity (Table 1 and Figure 1).

Rapid urbanization, drastic deforestation, human interference, rapid growth of automobiles was the factors that cause the adverse change of climatic condition in micro level. Irregular rainfall is the consequence of this change.

Table 1: Year and Month Wise Rainfall (in cm) -(2010 to 2015)

Month/Year	2010	2011	2012	2013	2014	2015
January	NIL	0.05	0.24	0.67	NIL	0.92
February	NIL	NIL	0.24	NIL	1.85	0.94
March	4.90	3.70	0.26	0.47	0.32	1.58
April	13.96	8.99	8.32	11.11	1.32	5.68
May	21.78	7.13	19.00	20.05	27.23	26.52
June	44.75	30.11	64.95	24.90	34.66	51.15
July	32.81	29.23	46.62	46.62	23.13	19.16
August	32.00	25.96	27.83	27.83	54.72	43.13
September	17.64	16.37	21.72	21.72	27.72	20.07
October	1.64	1.12	5.22	5.22	3.27	5.22
November	0.13	NIL	NIL	0.10	NIL	3.31
December	NIL	NIL	NIL	NIL	NIL	NIL

Source: Nimtijhora Tea Estate., Alipurduar, North Bengal

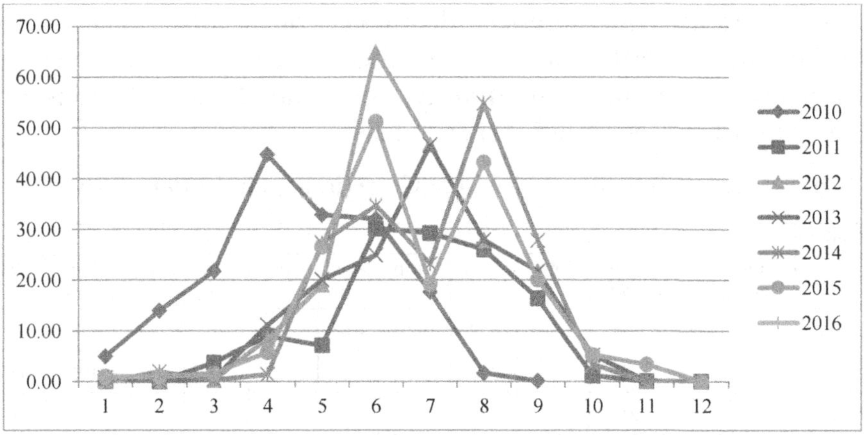

Figure 1: Rainfall Pattern in Dooars

Observation and Analysis I

From the Table 1 and Figure 1, it is revealed that (i) the region got winter rain in the year of 2011,2012,2013, and 2015 whereas in the year 2010 and 2014 experienced no rain fall in the month of January and February. (ii) In 2010 and 2011, the month of March received better rainfall in comparison to the next years. (iii) The highest rain fall was observed in the month of June in the years of 2010, 2011, 2012 and again in the year 2015, while in the year 2013 it was in the month of July and in the year 2014 it was in the month of August. (iv) In 2011 and 2012 June rain fall was found quite high in comparison to April and May rain fall of these consecutive years. (v) Moreover, in 2012 the June rain fall was so heavy within a few days that led to water logged condition in some lower parts of the field. (vi)May 2011 and April 2014 have shown huge reduction of rain fall. (vii) On the contrary, 2015 experienced drastic reduction of rain fall in the month of July after getting huge rainfall in the month of June. The year 2010, 2014 and 2015 experienced early monsoon, whereas 2011, 2012 and 2013 met the late monsoon.

The rainfall has shown significant irregularity even in the monsoon i.e. in the month of June, July and August and that irregular pattern continued even in September and October. The tea cultivation and allied agricultural practices (like pruning, administering of manures) depends significantly on the winter rain (November, December, January) and the rain of the month of April and May because the subsoil water level becomes less during these months. Every tea plantation unit would try to cop up with the water stress by the process of irrigation. The less rain in the month of April or May causes the reduced yield of the second flush of tea and reduced the quality. Also, the change of monsoonal pattern of rain causes significant crop loss.

Table 2: Rain Fall (Cm) and Yield of Crop (in Kg.) in Dooars, 2010-2015

Month	2010		2011		2012		2013		2014		2015	
	Rain fall	Crop	Rain fall	Crop	Rain fall	Crop	Rain fall	Crop	Rain fall	Crop	Rain fall	Crop
January	0	10336	0.05	21746	0.24	19194	0.67	0	0	0	0.92	0
February	0	358	0	741	0.24	1930	0	627	1.85	515	0.94	1873
Total	0	10694	0.05	22487	0.48	21124	0.67	627	1.85	515	1.86	1873
March	4.90	52242	3.70	38811	0.26	40506	0.47	41515	0.32	36698	1.58	52270
April	13.96	50114	8.99	23756	8.32	32630	11.11	32561	1.32	17250	5.68	36085

Month	2010		2011		2012		2013		2014		2015	
	Rain fall	Crop	Rain fall	Crop	Rain fall	Crop	Rain fall	Crop	Rain fall	Crop	Rain fall	Crop
Total	18.86	102356	12.69	62567	8.58	73136	11.58	74076	1.64	53948	7.26	88355
May	21.78	85513	7.13	54074	19.00	59583	20.05	53876	27.23	52545	26.52	68158
June	44.75	75875	30.11	27847	64.95	51794	24.90	49380	34.66	82118	51.15	55704
July	32.81	99469	29.23	74048	46.62	81759	46.62	91557	23.13	77561	19.16	56125
August	32.00	93226	25.96	60138	27.83	70059	27.83	76907	54.72	121909	43.13	76871
Total	131.34	354083	92.43	216107	158.40	263195	119.4	271720	139.74	334133	139.96	256858
September	17.64	105019	16.37	76252	21.72	70544	21.72	100510	27.72	101687	20.07	92740
October	1.64	132825	1.12	119584	5.22	75613	5.22	140645	3.27	133844	5.22	83810
November	0.13	99011	0	99574	0	63971	0.10	68502	0	92611	3.31	68695
Total	19.41	336855	17.49	295410	26.94	210128	27.04	309657	30.99	328142	28.6	245245
December	0	36223	0	33581	0	38282	0	65346	0	55799	0	59895
Total	0	36223	0	33581	0	38282	0	65346	0	55799	0	59895

Source: Nimtijhora Tea Estate., Alipurduar
[**Subtotal were mentioned to clarify the division of the year in Winter rain, pre-monsoon rain, Monsoon and post monsoon rain]

Observation and Analysis II

It is observed from the Table 2, that (i) in the year 2011 and 2012, the total production in the March and April was quite high even lack of winter rain because of the better rain fall in the month of March and April. (ii) But there is a clear indication that yields of crop and total production has depended up on the quantity/volume of rain fall especially at the beginning of the new flush. If the production of the month from May to August of the year 2014 and 2015 has taken into consideration, it is revealed that the production of monsoonal flush in the year 2014 i.e. from July went ahead to the year of 2015 because of the increased rainfall during the period.

Hence to overcome the stress of water due to lack of rainfall or irregular rainfall, the Tea production units have to set up the irrigation equipments (especially the sprinklers) and used to start irrigation from the end of November up to End of February or first week of March. But the environment has so far been changed that the irrigation needs to be continued even the last week of April. This causes huge expense that effects cost. This also leads to excess loss of underground water and of course excess use of fuels that lead to environmental degradation initially in the micro area.

Statistical Analysis to Check the Hypothesis

The month and year wise correlation has been checked at first, but for the convenience, the total rainfall and total yield of the years has been tabulated and correlation has been checked. [unit of rain fall in cm & crop in kg]

Table 3: Total Rainfall and yield of Crop in Dooars

Month	2010		2011		2012		2013		2014		2015	
	Rain fall	Crop	Rain fall	Crop	Rain fall	Crop	Rain fall	Crop	Rain fall	Crop	Rain fall	Crop
Jan	0	10336	0.05	21746	0.24	19194	0.67	0	0	0	0.92	0
Feb	0	358	0	741	0.24	1930	0	627	1.85	515	0.94	1873
March	4.9	52242	3.7	38811	0.26	40506	0.47	41515	0.32	36698	1.58	52270
April	13.96	50114	8.99	23756	8.32	32630	11.11	32561	1.32	17250	5.68	36085
May	21.78	85513	7.13	54074	19	59583	20.05	53876	27.23	52545	26.52	68158
June	44.75	75875	30.11	27847	64.95	51794	24.9	49380	34.66	82118	51.15	55704
July	32.81	99469	29.23	74048	46.62	81759	46.62	91557	23.13	77561	19.16	56125
Aug	32	93226	25.96	60138	27.83	70059	27.83	76907	54.72	121909	43.13	76871
Sep	17.64	105019	16.37	76252	21.72	70544	21.72	100510	27.72	101687	20.07	92740
Oct	1.64	132825	1.12	119584	5.22	75613	5.22	140645	3.27	133844	5.22	83810
Nov	0.13	99011	0	99574	0	63971	0.1	68502	0	92611	3.31	68695
Dec	0	36223	0	33581	0	38282	0	65346	0	55799	0	59895
Total	169.61	840211	122.66	630152	194.4	605865	158.69	721426	174.22	772537	177.68	652226

Table 4: Total Rainfall and Yield in Dooars

Year	Rain Fall (cm)	Yield (kg.)
2010	169.61	840211
2011	122.66	630152
2012	194.4	605865
2013	158.69	721426
2014	174.22	772537
2015	177.68	652226

Table 5: Correlation Check

	Rain fall	Yield
Rain fall	1	
Yield	0.066684806	1

Analysis & Inference

The value was 0.0667 which meant the correlation of those two variables were positive but weakly related. The cause of that value was the presence of other factors for the yield. The major factor was the prolonged use of irrigation process to compensate the reduced or irregular rain fall and maintain the yield. That irrigation compensated the soil water reserve on which the yield depended. That was the cause of the value while checking the correlation between the two variables, i.e. rain fall and yield. That prolonged irrigation proved the significance and necessity of rainfall with regularity.

Therefore, the Alternate hypothesis (H1) has been established.

Conclusion

The study concentrated exclusively on the matter of change of environment in a specific area and its significant affect on the micro climate. Thus the study has chosen the tea plantation of a specific area, to find if there was any significant effect on yield of crop within a five-year time frame (2010-2015 as the secondary data received from the source). The study revealed that the environment of the area has been changing rapidly due to different causes (especially deforestation, unplanned urbanization etc.) and that has rendered change in micro climate especially the rain fall in that area. The rain fall was found over all satisfactory as revealed from the data, but found quite irregular in the specific time frame that became critical in case of the yield of the Tea. That irregular rain fall led to the shortage of underground water especially in the sub soil level that might cause the water stress condition in the winter period. The condition has become so worse that prolonged irrigation would be required to compensate such water stress situation which proved that the yield was significantly dependent on the volume of rain in a regular pattern. So many other factors affecting the yield were revealed while studying the microclimatic level which was either natural or some social parameters. But, those were not taken into consideration in that study. The study ultimately reached to the conclusion that the microclimatic condition, even in terms of rain fall has been changed in this Dooars Region by showing its irregular pattern.

Acknowledgements

- Dibrugarh Company Ltd., Hasimara.
- Dutta Biswajit, Manager, Nimtijhora Tea Estate.
- Nimtijhora Tea Estate, Alipurduar

References

Ali Mohammed, Islam Mahmuda, Saha Narayan, Kanan Akber Hossain: *Effects of Microclimatic Parameters on Tea Leaf Production in Different Tea Estates in Bangladesh,* World Journal of Agricultural Sciences 10 (3): 134-140, 2014 [DOI: 10.5829/idosi.wjas.2014.10.3.1739]

Bhagat, R.M., Ahmed,K.Z., Gupta, N. & Baruah, R.D (India),. Wijeratne, M.A (Sri Lanka), Bore, John K. & Nyabundi, Darl W (Kenya) , Han,Wenyan, Li, Xin, Yan,Peng & Ahammed, G.J. (China): *Report of the Working Group on Climate Change of the FAO Intergovernmental Group on Tea* : A subsidiary body of the FAO Committee on Commodity Problems (CCP), Food and Agriculture Organization (FAO) of the United Nations : Rome, 2016.

Han, Wenyan, Li, Xin, Yan, Peng, Zhang, Liping: *Tea cultivation under changing climatic conditions,* Global tea science, February 2018, 455 - 472 [DOI: 10.19103/AS.2017.0036.19]

Othieno, C.O., W. Stephens and M.K.V. Carr, 1992. *Yield variability at the Tea Research Temperature and humidity had no direct effect on tea leaf foundation of Kenya. Agricultural and Forest production Meteorology*, 61: 237-252

Part - II

Environmental Challenges in the Present Context

7

Global Environmental Challenges in the 21st Century

Gopal Das[1] & Dr. Ranjan Sarkar[2]

[1]Assistant Teacher in Geography, Maynaguri Subhas Nagar High School, Maynaguri, Jalpaiguri, West Bengal, India

[2]Assistant Professor, Department of Geography, Ananda Chandra Training College, Jalpaiguri, West Bengal, India

Abstract

Global Environmental Challenges in 21st Century is the biggest concern in terms of Sustainability and accountability. These challenges mainly refer to those issues raised due to human activity and more emphasize has given to Worldwide Climatic change. Our earth environmental features have been altered during last few decades and likely to be altered during near future if we don't realize the essence of Environment on Biosphere. Human activity is primary and most prominent responsible factor for deteriorating the Environmental Balance and a lot of evidences may be summarized for this argument but at present century the foremost question is that what and how long we could realize the fact.

The present paper aims to study the basic Environmental Challenges of our earth. The study is based on environmental data of different International organization which were continuously engaged in world environmental management and preparedness. With the help

[1] Corresponding Author

of ArcGIS software version 10.3.1, different maps have been producing to show the continent wise and country wise environmental statistics.

The basic finding of this study stated that, our earth has experienced a drastic change in climate during last few decades and this is mainly due to anthropogenic factor. With the growth of world population and urbanization, the land use of different countries has changed to the direction of Natural to Artificial and this increased greenhouse gases in our environment. The changing land use pattern has also brought about a modification of natural habitat which accelerated the loss of biodiversity worldwide.

Keywords: Global Environmental Challenge, Environmental Sustainability, Land Use, Greenhouse Gas, Urbanization.

Introduction

The Environmental challenges simply refers to issues that should have the potentiality of alteration the course of life or normal pattern of life on this planet. These issues significantly brought up by human interference in to the nature. At the door of 21st century, the human welfare and Socioeconomic development become the prime objective for human survival throughout the world but the hidden worth for human existence is continuously remain abandoned. Human cannot survive without blessing of environment but the present century brings forth a lot of issues that have been raised due to excess human intervention in to the general system of nature. Global warming, toxic waste, water and air pollution, acid rain, emission of toxic gases, are frightening global environmental challenges that may threaten our future life as well as future generation if we do not perceive them. There are a number of factors among which the growth of world population and postmodern urbanization both are significantly attributed.

According to UNEP greenhouse gas emission report, global GHG emission grew at an average of 1.6 percent per year during the period from 2008 to 2017. The report in 2019 also indicates that global GHG emission continued to grow in 2018, breaking the 2017 record. According to UNEP, 2018 report China is the to greenhouse gas emitter followed by USA and India stands at fourth position. Many countries are vulnerable to climate change as well as biodiversity depletion. Global extraction of natural resources from ecosystem and mines grew more or less steadily over the past centuries. According to EEA report, during the 20th century, global sea level rose by an average of 1.7mm/per year and this was due to an increase rate in the volume of sea water as a consequence of global warming and melting of ice sheet of Greenland and Antarctica and this have been projected to rise continuously. The emission of CO_2 either due to man induced or due to natural incidence is directly impact on the global temperature balance and acidification of ocean water which cause the coral bleaching in tropical or sub-tropical region. This remarkably breakdown the symbiotic relation of different marine organisms and it may cause the destruction of marine ecosystem. Another most important thing to mention here that whole world is threatened by the changing patterns of global disease burdens in which we are currently suffering. So, a lot of issues in respect of global environmental perspective may be accounted but this is not enough until and unless we perceive those threaten and globally reform our environment by imposing sustainability which may firmly believe as a better way to forward.

The present paper may help to readers and policy makers for a competent perception about the current environmental challenges in global perspectives. This paper may also help to formulate some strategy and remedial measures for making our global environment healthy and worthwhile as may be forwarded wholesomely to our next generation.

Objectives of the Study

The fundamental objectives of this study are to show the major challenges in terms of global environmental outlook at 21st century. The present study also seeks to a spatio-temporal analysis of the environmental fact worldwide.

Materials and Methods

The present research is based on the information and data provided by various International Environmental organization for climatic change and contemporary environmental issues. Sentinel images have been used for understanding the spatial and temporal exposure of major environmental threaten. European Space Agency (ESA) is such an international agency, founded in 1975, which deals with space exploration and environmental management. Copernicus is the most ambitious Earth Observation programme to date. This programme produces an accurate, reliable and easily accessible information of environment to mitigate the impact of climate change and design a way for future reformation. The open access data of this programme is easy to download from Copernicus data hub. In this study a number of Sentinel 5 images have taken under deliberation. There are some National and International organization which were played an important role for executing and understanding the threats, promoted investigations and expert reports directing scientific research for identification of emerging problems and practical solution in long term basis.

International Environmental Organizations: United Nations Environment Programme (UNEP, New York), Global Green Growth Institute(GGGI,Seoul,SouthKorea),IntergovernmentalPanelonClimate Change (IPCC, UN, New York), International Union for Conservation of Nature (IUCN, Gland, Switzerland), European Environment agency (EEA), Copenhagen), partnership in Environmental Management for the seas of East Asia (PEMSEA, Quezon, Philippines).

Results and Discussion

Before starting the discussion, I solemnly remember a beautiful quote given by eminent Russian author Leo Tolstoy. *"One of the first condition of happiness is that the link between man and nature shall not be broken"* The link between man and nature is the central point to the debate on global environmental perspectives and the human activities were considered as a crucial factor which bring about some unprecedented environmental challenges and notable issues to contemporary society in 21st century. It has been supported by most of scientist and International body of Environment that the climate change may be considered as a component of unexpected environmental risk produced by technological advancement and globalization in current century. Exorbitant pollution from industrial activities, worldwide changing pattern of land use, spiralling population growth and overabundant urbanization leads to increase waste and toxic generation, poverty, overconsumption of natural resources creates a serious environmental issue which must be taken thoughtfully at the door of 21st century.

According to World Economic Forum 2007-2020, Global Risk Report, top five Global Risk in terms of Likelihood in 2020 is associating with Environmental aspect. Five most important risk factor are Extreme weather, Climate action failure, Natural Disaster, Biodiversity loss and human - made Environmental Disaster respectively. (World economic Forum, Global Risk Report 2020). According to this report the Climate Change is striking harder and more rapidly. The last five years are on track to be the warmest on record, natural disaster are becoming more intense and more frequent and last year witnessed unprecedented extreme weather throughout the world. Alarmingly, global temperatures are on track to increase by at least 3 degree C towards the end of the Century.

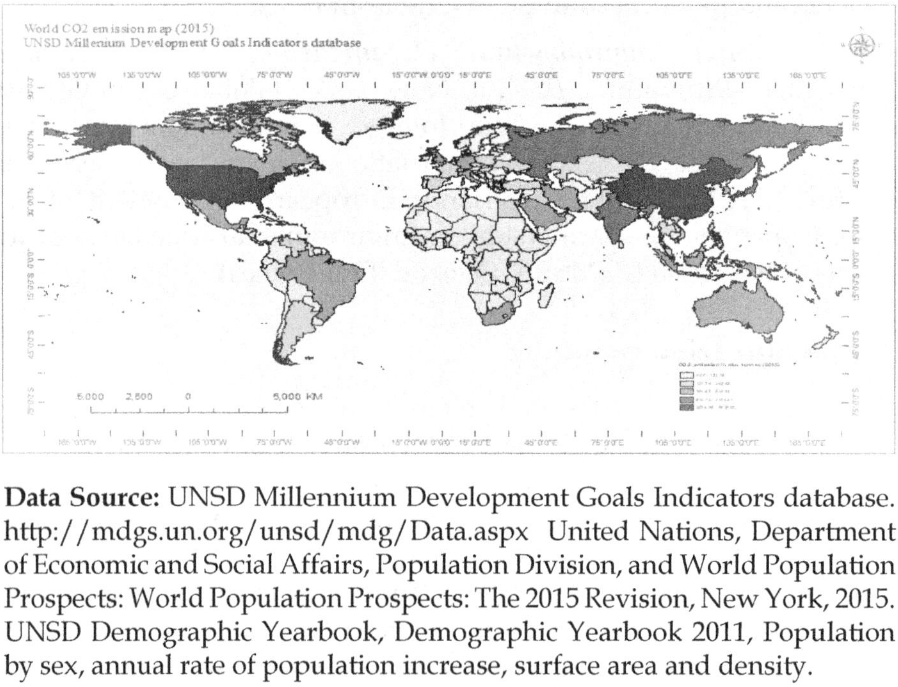

Data Source: UNSD Millennium Development Goals Indicators database. http://mdgs.un.org/unsd/mdg/Data.aspx United Nations, Department of Economic and Social Affairs, Population Division, and World Population Prospects: World Population Prospects: The 2015 Revision, New York, 2015. UNSD Demographic Yearbook, Demographic Yearbook 2011, Population by sex, annual rate of population increase, surface area and density.

Figure 1: World CO2 Emission

Measurement gathered by the ESA Copernicus sentinel -5P mission between April and September 2018 have been averaged to reveal nitrogen di oxide in the atmosphere. The main source of Nitrogen di oxide in atmosphere is the consumption of fossil fuels in mainly Industrial region. This pollutant harms the respiratory system. From the above figure it has exhibited that, most of south East Asian Countries were recorded a highest level of this pollutant. China, Indian subcontinent, Germany, France, Angola contribution a large amount of Nitrogen di oxide to global atmosphere whereas South America which is characterized by a large forested area (FAO, 2010) is less contributed NO2 to the Global atmosphere.

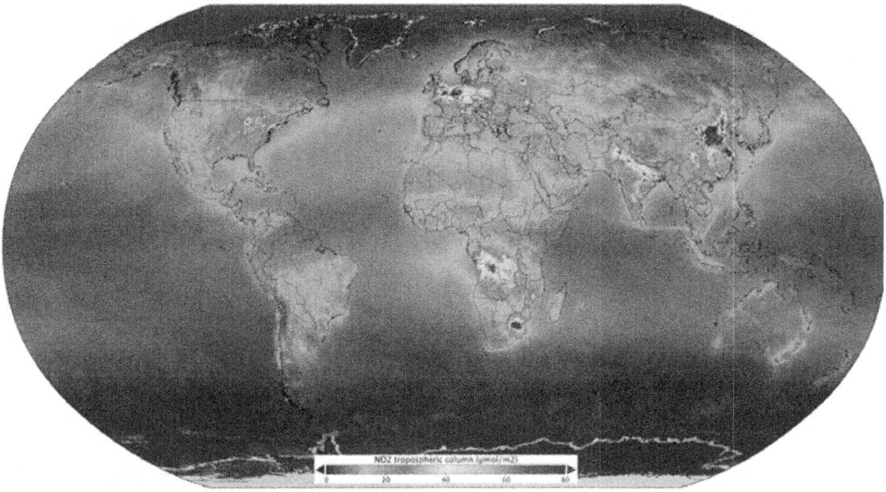

Image Source: European Space Agency. (https://www.esa.int/ESA_Multimedia/Missions/Sentinel-5P/(result_type)/images)

Figure 2: Worldwide Emission of Nitrogen Dioxide

The increasing rate of Global Temperature is another crucial matter of concern and the temperature trend is positively correlated with emission of greenhouse gases. From above figure it may be defined that most of South East Asian Countries produce a large quantity CO2. According to 2015 data China stands first in terms of CO2 emission followed by United States of America, India, Russia and Japan. These are the leading countries for generating CO2.

The figure 3 shows the leading PM2.5 concentration countries. PM2.5 is the pollutant which widely regarded as most harmful to human health. PM2.5 is defined as ambient air born particles measuring up to 2.5 micron in size. This microscopic size allows the particles to enter the blood stream via the respiratory system and travel throughout the body causing far reaching health effects including asthma, lung cancer, and heart disease. US AQI is among the most widely recognized index for assessing air quality. This Index converts pollution concentration into a colour-coded scale of 0-500, where higher values indicate increased health risk. The US AQI "good" range (<12μg/m^3) is marginally higher than the WHO air quality guideline (<10µg/m³) (2019 World Air Quality Report, Region and City PM2.5 Ranking, IQ Air).

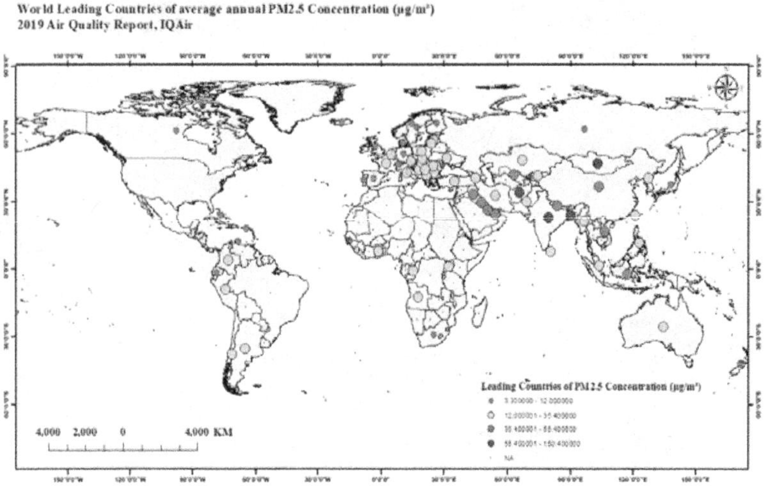

Figure 3: World Leading Countries of Average Annual PM2.5 Concentration, 2019

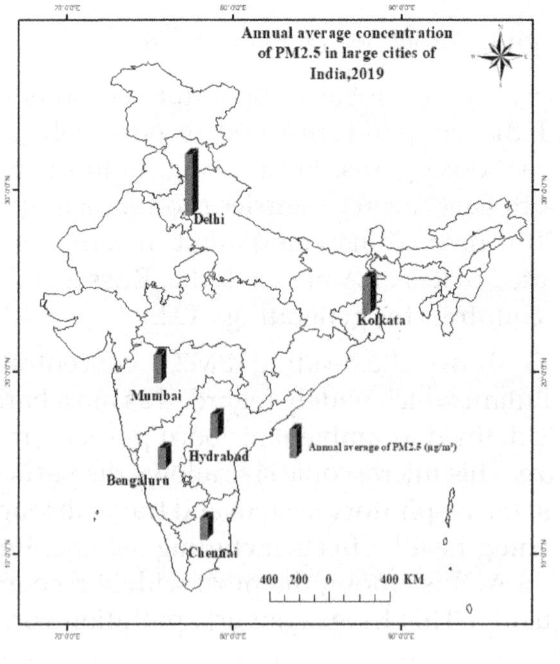

Figure 4: Annual Average Concentration of PM2.5 in Large Cities of India, 2019

According to US Air Quality Index four regional capital cities accompanied with the higher level of annual average PM2.5 concentration. These four capital cities are Delhi from India, Dhaka from Bangladesh, Ulaanbaatar from Mongolia, and Kabul from Afghanistan.

According to data the capital city of India is facing a serious challenge in respect of PM2.5 concentration level followed by Kolkata, Mumbai. India has insufficient air quality monitoring network to its population size.

Major Environmental Issues at Global Level

At state of the art the people oriented development in both developed and developing countries have intricated the man and nature relasionship to a substantial extent. The rapid growth of population coupled with technological advancement exaggerated a number of concerns which directly strike the environmental sustainability at global scale. The major isuues may have summerized in the following way:

- Depletion of Natural Resources.
- Air pollution.
- Water Pollution.
- Shortage of ground water storage and movement.
- Ground water pollution.
- Loss of fertility of soil or soil pollution.
- Ozone layer depletion.
- Global warming.
- Melting of ice sheet in polar areas and rise in sea level.
- Bio-diversity loss.
- Extinction of wildlife and loss of habitat.
- Outbreak of communicable disease or health emergency.

It is very important to note that there is a strong linkage among different challenges by some or the other ways but a very little could identify such linkage and it is also very difficult to tackle challenges one by one. Every component of our environment must be apprised with the real time observation and a comprehensive and collective measures should be taken under compilation.

Depletion of Natural Resources

The depletion of Natural Resource is one of the fundamental issues for global challenge. Different analysis and research have shown that the poor countries or the developing countries are losing natural resources fast without gaining much by other alternatives. Over extraction of minerals as well as other non-renewable resource in large scale arise the question mark towards how much wealth a country could accumulate for its future generation.

According to Statistical Review of World Energy, 2020 (69th edition) (www.bp.com/statisticalreview) Global coal production rose by 1.5%, with China and Indonesia providing the only significant increase and the largest declines came from US and Germany. Oil consumption grew by 0.9 million barrels per day but the decline of OPEC oil production by 2 million barrel per day and this is the largest decline since 2009.

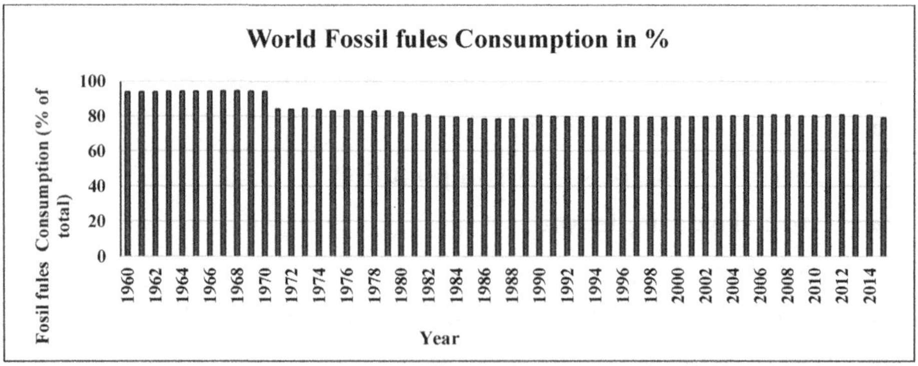

Data Source: Fossil fuel comprises coal, oil, petroleum, and natural gas products. IEA Statistics © OECD/IEA 2014 (http://www.iea.org/stats/index.asp), subject to https://www.iea.org/t&c/termsandconditions/

Figure 5: World Fossil Fuels Consumptions in Percentage (%) from 1960 to 2014

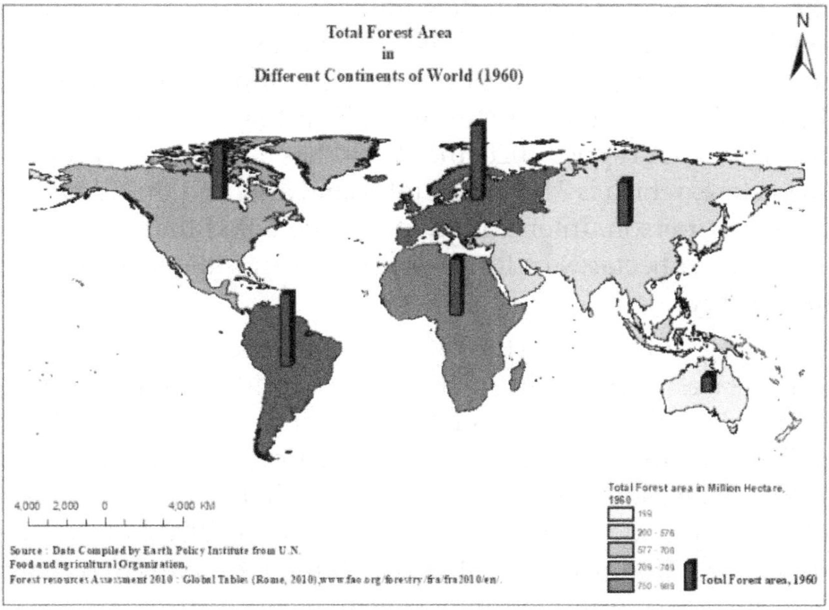

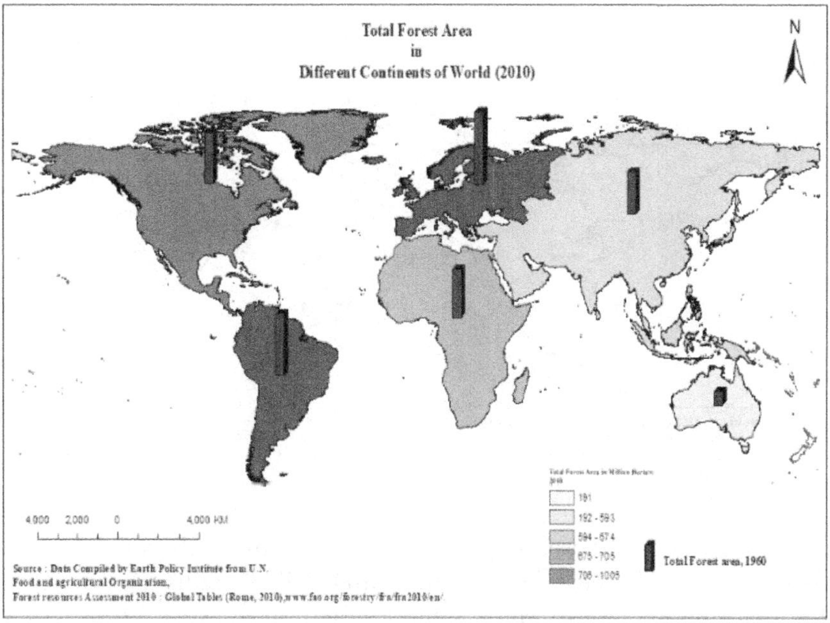

Source: Data Compiled by Earth Policy Institute from U.N. Food and agricultural Organization, Forest resources Assessment 2010: Global Tables (Rome, 2010), www.fao.org/forestry/fra/fra2010/en/.

Figure 6: Continent Wise Total Forest Area in Million Hectare (1960 And 2010)

The forest resource is another influential part for environmental sustainability at Global as well as regional level. The Global Forest Resource Assessment (FRA 2020) provide a comprehensive view of the World's forest and the ways through which the resource is changing. According to FRA report, 2020, the World has a total forest area of 4.06 billion hectare, which is 31 % of the total land area and this is equivalent to 0.52 ha per person. Interestingly it may be noted that the world has lost 178 million hectares of forest since 1990 and this is an area about the size of Libya.

According to FRA report, 2010 total forest area of World was 4168 million hectares in 1990 which has reduced to 4033 million hectares in 2010. Africa had the largest annual rate of net forest loss in 2010-2020 at 3.9 million hectares, followed by South America, at 2.6 million hectares. According to an estimate of FRA, 2020 420 million hectares of forest has been lost worldwide through deforestation since 1990 but the annual rate of deforestation was estimated at 10 million hectares in recent time (2015-2020), down from 12 million hectare in 2010-2015).

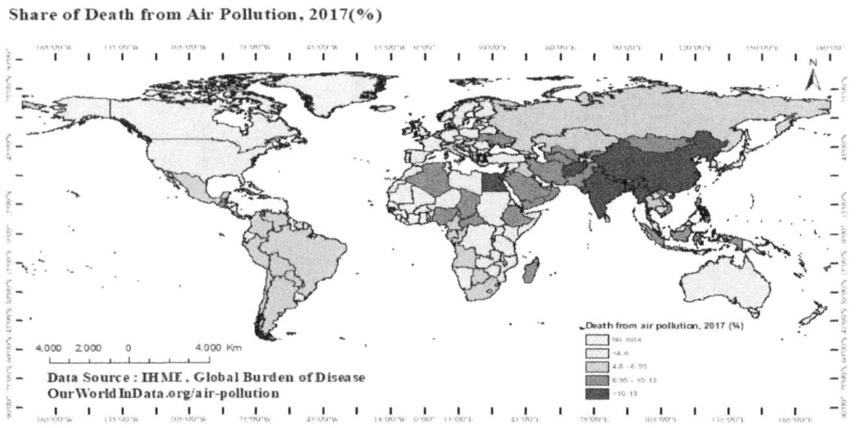

Source: Data Published by Global Burden of Disease Collaborative Network. Global Burden of Disease Study 2017 (GBD 2017) Results. Seattle, United States: Institute for Health Metrics and Evaluation (IHME), 2018.

Figure 7: Country Wise Share of Death from Air Pollution (%), 2017

Air Pollution: Air pollution is another important issue which has been linked with a lot of environmental phenomenon and is determined by the concentration of a number of toxic gases mainly released from anthropogenic structures or human activity. Major pollutants are Co2, O2, No2, So2 etc. All these pollutants had a great impact on human health either short- or long-term base. The concentration level of different pollutant in atmosphere has increased at global scale and this is responsible for global temperature rise. There are a lot of evidence of O3 injury to trees in Greece, Italy (Busotti and Ferretti 1998). According to Our World Database air pollution is attributed to 9% of death globally.

From figure 7 it may be stated that the most of South East countries have recorded highest percentage of deaths from air pollution. According to the report South Korea has the highest percentage i.e. 16.50 in 2017, which has falling down from 17.19% in 1990. India has a percentage of 12.52 in 2017, rising up from 12.15 in 1990. The percent share of death for whole world was estimated as 10.10% in 1990 but the share has decreased to 8.75% in 2017(IHME, 2018).

Water Pollution and Ground Water Storage: According to UNDESA, 2017 the World population is projected to reach 9.8 billion people in 2050.The rapid population growth and changing pattern in consumption will require more production in agriculture. This will be led to a new environmental externality including impact on water quality. The intensification of agriculture or the productivity requires developed irrigation facility which somehow increase the salinity by mobilizing salts accumulated in the soil and this results into the high salinity of grown water or surface water. This salinization can alter the geochemical cycle or effects on the biodiversity of microorganisms. More extraction of ground water significantly influences on the storage and discharge of ground water. According to FAO 2011, Argentina, Australia, China, India, United States of America and many countries of central Asia are suffering from the problems of saline water and according to IGRAC, 2009 approximately 1.1 billion people lived in regions that had saline ground water. Organic substance from animal excreta or food are another major source of water pollutant which reduce the oxygen level of water. According to SACEP, 2014 shrimp aquaculture in Bangladesh generates 600 tonnes of waste per day in to water. In Scotland a large quantity of organic waste had produce from salmon production. Rapid growth of population and resultant urbanization extinguish a number of reservoir or wetland in urban areas this may be also considered as a serious problem towards the way of water resource management in urban landscape.

Ozone Layer Depletion: The ozone layer depletion is one of the very important environmental challenge at current century. According to NASA the ozone hole is not technically a "hole" where no ozone is present, but is actually a region of exceptionally depleted ozone in the stratosphere over the Antarctic that happens at the beginning of Southern Hemisphere spring (August -September). The ozone layer protects us from most damaging ultraviolet rays coming from sun. In 1974, Scientists discovered that chlorine and bromine containing compounds such as chlorofluorocarbons (CFCs) and halons that could deplete this protective layer and by 1980s, they had observational evidence that it was happening. The Montreal Protocol was signed on September 16, 1987 for protecting this planet from such environmental problems. The ozone hole is the most recent phenomenon which is directly affected by human activity. The ozone layer depletion has become more than obvious over the past decades.

The consumption of ozone depleting substance was much higher in earlier decades. According to UNEP ozone depleting substance are any substance containing chlorine or bromine that destroy the stratospheric ozone layer.

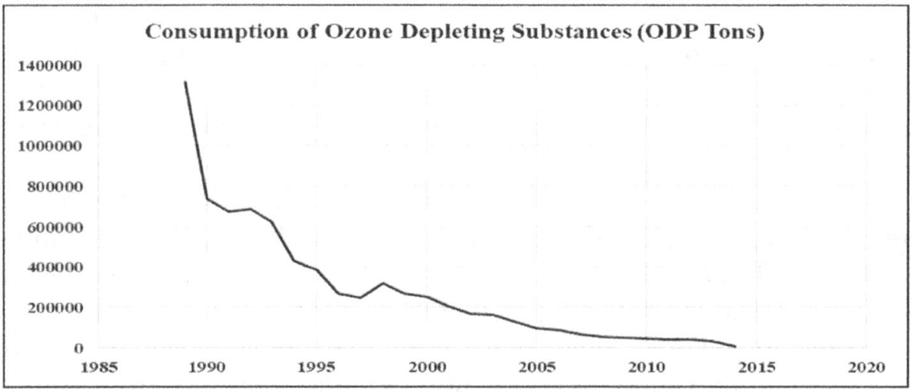

Data Source:UNEP(2020): The UNEP Environmental data Explorere, as compiled from United Nation Environment programme.http://ede.grid.unep.ch.

Figure 8: Global Consumption Ozone Depleting Substances (1989-2014)

During 1989 to 1990 the Global consumption of ozone depleting substances was very high but the amount become low during present time. After Internation aggrement about Ozone hole Protection, the

Global emission of Ozone depleting Substances have declined by 98%. According to ourworlddata report, 2014 totat world ODS was recorded as 6922 tonnes.

The excess emision of ODS and resultant ozone depletion cause of increase UV radiation reaching the earth surface which can pose a number of health effects likeskin cancer, accerliation of eye cataracts development, reduction of immune system, loss of biodiversity etc.

Global Warming: Global warming is the rapid increases in earth temperature due to release of greenhouse gases through over uses of fossil fuels. This is the another problem which born up some significant environmental issues. The earth surface and atmospheric temperature has increased in unusual manner during last few decades. At the present century the surface temperature is a great issue which drastically impacted on polar ice cover and sea level changes.

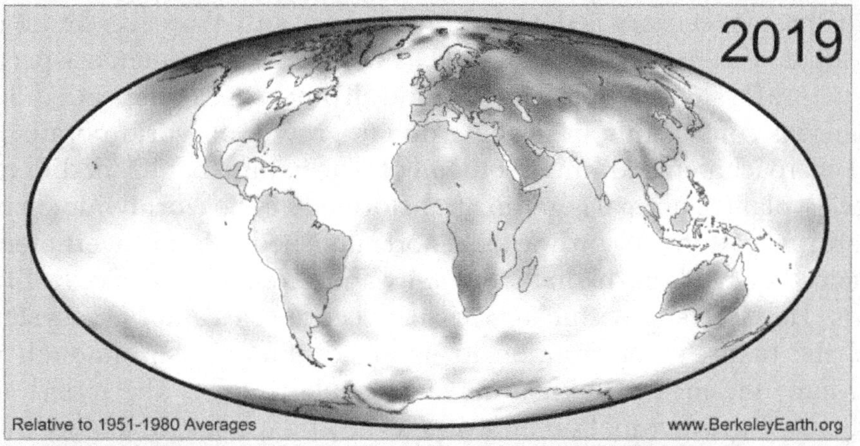

Source: www.BerkeleyEarth.org In the map shown above taken from the Berkeley Earth global temperature Report and downloaded from our world in data platform, indicates the global distribution of temperature Changes in 2019 relative to the period 1951-1980. (http://berkeleyearth.org/archive/2019-temperatures/)

Figure 9: World Temperature Anomaly (0 C), 2019

The report also indicates that the global temperature over the land surface has increased around twice as much as the ocean and this is responsible for the higher change in average temperature in northern hemisphere than south. According to our world in data report the global average temperature anomaly has been recorded as 0.78°C for upper, 0.70 °C for lower and 0.74 °C for median whereas it was -0.15 °C as upper, -0.20 °C for median and -0.27 °C for lower in 1900. The change of global average temperature is directly dependent on the change of emission of different greenhouse gases like CO2, NO2, and CH4 etc. The biggest impact of global warming is confined in polar areas where the polar ice cap is gradually melting and it results into the increase of sea level. The gradual warming of this planet is responsible for coral bleaching which eventually destroy its biodiversity. Different research and study opined that the melting of ice cover in polar areas can generate different unknown germs which could brought some global pandemic situation. So, there are a number of linkages with global warming and another environmental challenges and its magnitude of occurrence.

Loss of Biodiversity: The ecosystems and global ecological balance at the present century is deteriorating very rapidly than ever and there are lot of evidences. The ecological balance is the fundamental aspect of the natural system on which human and all other species depended. This balance is consistently destroying due to growth of world population and encroachment of human settlement. The natural habitat of different species, plant or animal is getting transformed for the worldwide change in land use pattern. According to a report published by Intergovernmental Science-Policy Platform on Biodiversity and Ecosystem Service (IPBES), Paris 'The diversity within species, between species and of ecosystems and the nature's contribution to us are declining fast, although we still have the means to ensure a sustainable future". The report also elaborated the fact that on an average 25% of species threatened with extinction across terrestrial, freshwater and marine vertebrate, in vertebrate and plant groups that have already been studied in sufficient detail. (http://www.un.org/sustainabledevelopment/blog/2019/05/nature-decline-unprecedented-report/

The rising surface temperature of oceanic water and other factors caused a destruction of coral reefs in form of coral bleaching. According to National Oceanic and Atmospheric Administration (NOAA), mass coral bleaching events have been occurring more and more frequently in last 30 years. A large number of corals in United States have been bleached. The north western Hawaiian Island, Marshall Island had experienced sever bleaching of corals due to record heat stress during third global coral bleaching event which was started in mid-2014 and continue up to May, 2017. The above figure shows the global hotspot regions of (https://coralreefwatch.noaa.gov/data/5km/v3.1/current/daily/gif/cur_coraltemp5km_hs_ large.) coral bleaching which is more prominent in north eastern pacific, north-western pacific and north-western Atlantic Ocean.

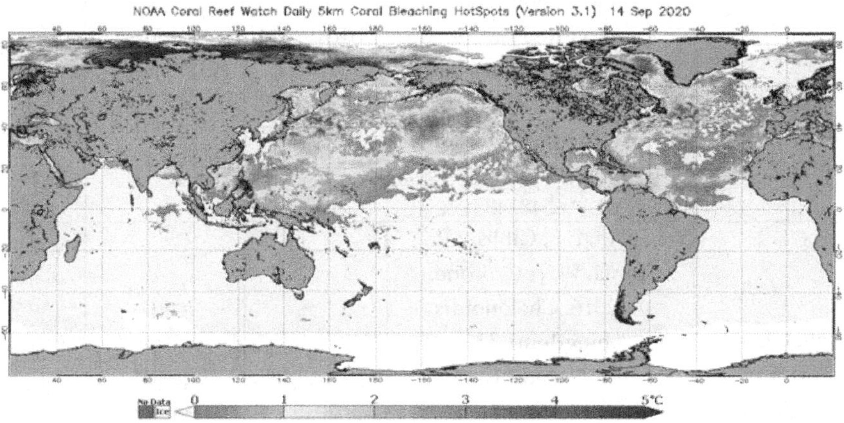

Figure 10: Global Coral Bleaching Event (NOAA, 2020)

The biodiversity conservation is a very crucial part in respect of environmental as well as food and agricultural concern. The 'associated biodiversity" is referring to the vast range of organisms that live in and around food and agricultural production system, sustaining them and contributing to their input. This is also important for suitable development. To meet the basic requirement of growing population the food and agriculture is indispensable for food security and a lot of organisms in soil sustain the food chain or maintain the linkage between various tropic levels. So, the biodiversity provides all basic elements to sustainable growth of agriculture and food security.

Extinction of Wildlife: The International Union for Conservation of Nature (IUCN) which officially known as International Union for Conservation of Nature and Natural Resources and Formerly known as International Union for the protection of Nature is an International organization with headquarter in Gland, Switzerland focus on the compilation and publication of Red list of Threatened species, and assess the conservation status of biological species worldwide. As per the IUCN Report some extinct species and their assessment has furnished in table 1.

Table 1: Some Extricated Species, IUCN

Species name	Citation	Last Assessed	Country occurrence	Status
Splendid poison frog (oophaga speciosa)	Cayot, L.J., Gibbs, J.P., Tapia, W. & Caccone, A. 2016. Chelonoidis abingdonii. The IUCN Red List of Threatened Species 2016: e.T9017A65487433. https://dx.doi.org/10.2305/IUCN.UK.2016-1.RLTS.T9017A65487433.en.	11th august, 2015	Equador (Galápagos)	Extinct
Pinta Giant Tortoise (Chelonoidis abingdonii)	Cayot, L.J., Gibbs, J.P., Tapia, W. & Caccone, A. 2016. Chelonoidis abingdonii. The IUCN Red List of Threatened Species 2016: e.T9017A65487433. https://dx.doi.org/10.2305/IUCN.UK.2016-1.RLTS.T9017A65487433.en.	11th august, 2015	Equador (Galápagos)	Extinct
Christmas Island Pipistrelle (pipistrellus murrayi)	Lumsden, L., Racey, P.A. & Hutson, A.M. 2017. Pipistrellus murrayi. The IUCN Red List of Threatened Species 2017: e.T136769A518894. https://dx.doi.org/10.2305/IUCN.UK.2017-2.RLTS.T136769A518894.en	7th August, 2016	Christmas Island, Australia	Extinct
Chatham Bellbird	BirdLife International. 2016. Anthornis melanocephala. The IUCN Red List of Threatened Species 2016: e.T22728814A94997726. https://dx.doi.org/10.2305/IUCN.UK.2016-3.RLTS.T22728814A94997726.en.	1st October, 2016	New Zeland	Extinct

Species name	Citation	Last Assessed	Country occurrence	Status
Dodo (Raphus cucullatus)	BirdLife International. 2016. Raphus cucullatus. The IUCN Red List of Threatened Species 2016: e.T22690059A93259513. https://dx.doi.org/10.2305/IUCN.UK.2016-3.RLTS.T22690059A93259513.en.	1st October, 2016	Mauritius	Extinct
Morant's Blue (Lepido-chrysops hypopolia)	Cockburn, K.N.A. 2020. Lepidochrysops hypopolia. The IUCN Red List of Threatened Species 2020: e.T11537A168301541. https://dx.doi.org/10.2305/IUCN.UK.2020-2.RLTS.T11537A168301541.en.	20th February, 2018	South Africa (Kwazulu-Natal)	Extinct

Conclusion

At the state of 21st century, the various environmental phenomenon that can be said to contribute to the rubric include global warming, climatic change, ozone depletion, biodiversity loss over space and time. These environmental extremities have a drastic effect on human health and also on whole biosphere of this planet. There are still much that is unknown about the potentiality of such problems. The growth of population and urbanization as well as change in land use pattern escalating the whole situation. Time has now come to understand essence of Environmental Sustainability to ensure a betterment of the Natural System and Ecological balance for our future generation. Comprehensive strategies and measures should be taken form ground level survey. A lot of issues are related to the climate change and deforestation. Different worldwide organization are working in this direction, that will help to offset our carbon footprint which is a measure of the impact of our activities have on the amount of carbon dioxide (CO2) produced through the burning of fossil fuels and is expressed as a weight of CO2 emission produced in tonnes (WHO).So finally it may be conclude that the problem is associated with long term and somehow with short term effect which must be focussed with the lens of cognizance to flourish a new way to forward.

Acknowledgement

Author extend his heartfelt thanks and gratitude to all personals and officials of different National and International organizations; whose countless effort make him possible to extract different data and information from various online sources.

References

Anand SV (2013) Global environmental Issues.2: 632 doi:10.4172/scientificreports.632

Bussotti, Filippo & Gerosa, Giacomo & Cozzi, Alberto & Ferretti, Marco & Bettini, Davide & Nibbi, Renzo. (2003). Crown Condition Surveys in Italian Forests: Issues in Reporting Findings. Environmental monitoring and assessment. 85. 221-38. 10.1023/a: 1023978614554.

Da Costa Ferreira, L., Barbi, F. (2016). The Challenge of Global Environmental Change in the Anthropocene: An Analysis of Brazil and China. Chin. Polit. Sci. Rev. 1, 685- 697.

Earth Policy Institute from U.N. Food and agricultural Organization, Forest resources Assessment 2010: Global Tables (Rome, 2010).

Eminer, Fehiman & Şafakli, Okan. (2014). A RESEARCH ON THE ENVIRONMENTAL PROBLEMS OF NORTHERN CYPRUS. Journal of environmental protection and ecology. 15. 468-477.

F. Bussotti, M. Ferretti. (1998). Air pollution, forest condition and forest decline in Southern Europe: An overview, Environmental Pollution, Volume 101, Issue 1,1998, Pages 49- 65, ISSN0269-7491, https://doi.org/10.1016/S02697491(98)000396.

(http://www.sciencedirect.com/science/article/pii/S0269749198000396)

Global Burden of Disease Collaborative Network. Global Burden of Disease Study 2017 (GBD 2017) Results. Seattle, United States: Institute for Health Metrics and Evaluation (IHME), 2018.

Global Forest resource Assessment, 2020, Key findings, food and agricultural Organization of the United Nations Global resource Outlook, 2019.

Harper C, Snowden M (Eds). (2017). Environment and Society. Human Perspectives and Environmental Issues. 6th edition. Routledge, New York, London.

Jorgenson, Andrew & Dunlap, Riley. (2012). Environmental Problems. The Wiley- Blackwell Encyclopaedia of Globalization. 10.1002/9780470670590.wbeog174.

Khan, Mehran & Chang, Yen-Chiang. (2018). Environmental Challenges and Current Practices in China-A Thorough Analysis. Sustainability. 10. 2547. 10.3390/su10072547.

Saha, Sarmistha & Bindu, Samik & Chakrabarti Basu, Priyadarshini & Mukherjee, Tilottama & Mukherjee, Susmita & Bhattacharya, Nivedita & Selim, Abdus & Sarkar, Sayantan & Dasgupta, Shyamasree. (2019). Environmental Issues: Approaches and Practices.

Quesada-Ruiz, Samuel & Attie, J.L & Lahoz, William & Abida, Rachid & Ricaud, Philippe & El Amraoui, Laaziz & Zbinden, R. & Piacentini, A. & Joly, Mathieu & Eskes, Henk & Segers, A.J. & Curier, Lyana & Dehaan, Johan & Kujanpää, Jukka & Oude-Nijhuis, Albert & Tamminen, Johanna & Timmermans, Renske & Veefkind, Pepijn. (2019). Benefit of ozone observations from Sentinel-5P and future Sentinel-4 missions on tropospheric composition. Atmospheric Measurement Techniques Discussions. 1-34. 10.5194/amt-2018-456.

UN environment Annual report, empowering people to protect the planet, 2016, UN environment.

UNSD Millennium Development Goals Indicators database 2019 World Air Quality report, Region and City PM2.5 Ranking, IQ Air Statistical Review of World Energy, 2020 (69th edition).

Valavanidis, Athanasios. (2019). Current Environmental Issues and Emerging Global Challenges in the 21st Century for Environmental Protection and Sustainable Development. 1. 1-52.

Web References

https://www.unenvironment.org/resources/emissions-gap-report-2019 accessed on 10.09.2020 at 8.30pm IST.

https://www.eea.europa.eu/data-and-maps/indicators/sea-level-rise-5/assessment accessed on Sep. 15 2020 02:20:45 GMT+0530 (India Standard Time).

http://mdgs.un.org/unsd/mdg/Data.aspx accessed on Sep. 04 2020 02:05:35 GMT+0530 (India Standard Time).

https://www.esa.int/ESA_Multimedia/Missions/Sentinel-5P/(result_type)/images on Sep. 17 2020 01:25:55 GMT+0530 (India Standard Time).

http://www.bp.com/statisticalreview accessed on Sep. 05 2020 04:07:31 GMT+0530 (India Standard Time).

www.worldbank.org acceded on Sep. 05 2020 04:05:02 GMT+0530 (India Standard Time).

https://ourworldindata.org/accessed on Sep. 15 2020 02:20:45 GMT+0530 (India Standard Time).

8

Covid-19 Crisis Through A Migration Lens: Livelihood Effects for Pandemic Induced Displacement

Chandra Kanta Saha
Assistant Professor and Head, Department of Economics,
Sister Nibedita Government General Degree College for Girls,
Kolkata, West Bengal, India

Abstract

Migration studies are the oldest system of analysis of voluntary human spatial mobility. But in recent years involuntary mobility which is also known as forced migration in terms of internal displacement are much more pronounced in the context of migration studies. Different terminologies of displacement are coined like development induced displacement, conflict induced displacement, disaster induced displacement etc. Now the involuntary displacement of migrant workers induced by Covid-19 from destination to source has brought to fore the pandemic induced displacement. The economic crisis induced by Covid -19 could be long, deep and pervasive when viewed through a migration lens. Thy long-lasting lockdown has inadvertently exposed the extreme vulnerability of migrant workers in India. Migrants face the risk of contagion and also the possible loss of employment, wages which has a destabilizing effect on their livelihood. Governments need to address the challenges facing internal migrants by providing a mixture of in-kind transfer and cash transfer, by adopting universalization of PDS, by supporting NREGA program and its introduction in urban area and by providing one nation one ration card so that national portability will be achieved. Despite government promises and schemes to generate employment, a large number of migrants were returning to cities and the issue of re-migration arises. This paper also highlights this apprehension of the deficit of trust building among the migrants and it recommends the stimulus package of direct spending rather than increasing liquidity in the economy. This crisis compels us to rethink urbanisation that is more inclusive and sustainable and it is the long-term solution to ensure life with dignity.

Keywords: Pandemic Induced Displacement, Covid-19, Internal Migrants, Livelihood, Inclusive.

Introduction

The term 'displaced persons' was coined by sociologist M. Kulischer and applies this term to all categories of forced mobility in Europe during the war. Among all types of displacement, the problem of development induced displacement emerges as the world's largest category of internal displacement and it has a long-term impact. Development induced displacement and resettlement (DIDR) occurs when people are forced to leave their homes or land as a result of development. This is the concept of forced displacement which is involuntary in nature. It is related with construction of dams for hydroelectric power and irrigation, development of various projects like mining, industrial plants, railways, road developments, urbanization and forestry etc. This displacement is inherent in the policy of development which is guided by neoliberal economic reforms under globalization. In this context resettlement has an important role to play which refers to physical relocation combined with appropriate support mechanisms including social support in the new location.

Similar to this line of thinking, the unprecedented displacement crisis in this Covid-19 pandemic situation is known as pandemic induced displacement. Migration is a reality where workers are migrated to cities for better jobs and better income and to fulfill their aspirations and hope but Covid-19 situation forced them to flee again to their villages. This reverse migration can be treated as displacement and they are also known as internally displaced persons. This displacement from cities to villages has deepened their job and livelihood crisis and has also increased their vulnerabilities which they are not ready to cope up. In this paper an attempt has been made to analyze the livelihood effects of pandemic induced displacement and also the inadequacy of welfare measures in India to reduce the plight of the displaces. An important task undertaken by this paper is the analysis of long-term government measures which is based on correct planning of urbanization and a job guarantee program for urban India. In this case the issue of resettlement implies the support mechanism for protecting livelihood, creation of sustainable employment and providing social safety nets for the displaces so that they can be able to maintain their livelihood in their source place after returning from cities in this crisis period. In long term sustainable policies will reduce the rate of migration and also reduce the vulnerabilities of the migrant workers and proper protection (in this case protection is used instead of resettlement) can transform the entire program into a positive sum game and has a potential for development. This humanitarian crisis could have been averted by better preparedness, setting up nationwide monitoring mechanisms and providing social and economic safety nets to the displaces.

Rationale of the Study

Both types of displacement (development and pandemic) have triggered the number of internally displaced persons in the economy. Two views are predominant in the analysis of displacement. One is physical and the other is economic. Physical displacement involves dislocation from one's habitat. It also has the impact on their livelihood which is known as economic displacement. Livelihood restoration is common objective of the government for both physical and economic displacement. In pandemic induced displacement we are concerned with economic displacement that is loss of livelihood and income and it is imperative to take care of the issues of vulnerability and marginalization of these displaces. In development induced displacement the issue of reestablishment of social network is also important along with the improvement in living standard. In pandemic induced displacement we are only concerned with livelihood analysis and there is no need of physical relocation as the migrants are switching from destination to sources. The common thing is that both of these can be treated as a distinct, negative phenomenon related to violation of fundamental human rights and ultimate aim of the study should be the expansion of individual and collective freedom by averting these types of forced displacement.

Pandemic induced displacement has a heavy bearing on livelihoods of displaces as these seasonal migrants belong to the socio economically weaker sections and based their livelihood on daily wage. The nationwide lockdown to combat the pandemic situation terribly impacted the unorganized sector where the employment of the migrants is largely contractual. It is true that they had built the cities with their sweat, hard work and toil but in this situation, they are forced to move from destination to source and some are trapped in the destination cities (due to lack of transport facilities) facing harsh conditions on their live and income and they are known as stranded migrants. Such a huge quantum of reverse migration has not been witnessed in our times. It is the challenge for the government to settle a meaningful solution for this tradeoff between live and livelihoods. It is a new type of displacement problem that world has ever seen.

Literature Review

According to Lewis model rural to urban migration is always beneficial as surplus laborers of agricultural sector are always absorbed in the modern industrial sector. This type of migration was deemed socially beneficial as resources were shifted from the place where social marginal product is zero to the place where it is positive.

According to Harris- Todaro model, recent LDC experience shows that this rate of rural urban migration continues to exceed rates of urban job creation and finally migrants are absorbed in informal sector of the urban economy. So, in these two models' migration is stimulated primarily by rational economic considerations of relative benefits and costs, mostly financial.

The problem of forced migration is very clearly depicted by Cernea (2000) in his Impoverishment Risks and Reconstruction (IRR) model. He summarized nine social risk affecting displaced people: landlessness, joblessness, homelessness, marginalization, food insecurity, increasing morbidity, loss of access to common property resources, education loss, and social disarticulation. These are also common for pandemic induced displacement.

Ray and Subramanian (May 2020) found that vulnerability to a lockdown is accentuated in India by three specific features of the population. The dominancy of casual labor, preponderance of informal production (these activities cannot be easily taken online) and low median household savings are the key features that has exacerbated the crisis of reverse migration. They have emphasized a redistributive plan and it is imperative to make transfer to protect the vulnerable. They have also raised the debate of lives versus lives, not a question of live versus economics and it is a question of which lives have greater visibility.

Azim Premji University partnering with ten Civil Society Organisations has conducted a survey on Covid-19 livelihood effects on different states of India. The report has admitted that the national lockdown has had a profound effect on employment and earnings. A survey (5000 respondents) from Azim Premji University across 12 major states over a month period between 13th April 2020 and 20th May 2020 revealed that 66% workers lost their employment, 77% households consuming less food than before and 47% households do not have enough money to buy even a week's worth of essentials. The worst hits are casual workers and self employed in both urban and rural areas. This shows the gravity of the problem with the sudden and severe lockdown. In this context the migrant workers have been the worst affected. The current relief level, especially direct public spending does not appear to be in commensurate with the severity of the situation and announced relief measures had not reached large sections of the economically vulnerable population. The unique contribution of the report was the proposal for the creation of a National Urban Employment Guarantee Programme for sustainable employment.

The movement from rural to urban is not the displacement issue as migrants move at their will. But in this pandemic situation reverse migration has occurred as they are forced to move from urban centers to village again. This can be seen as reverse exodus from migration point of view. These migrants were not considered as part of urban community by the urban development and planning, so management of these migrants under lockdown represent a massive logistic challenge for the government and this type of displacement has to be tackled as an emergency basis so that we can reduce the plights of the migrants. In this case livelihood displacement has a heavy toll on access to healthcare, food and nutritional intake and education of children.

Study Objectives

1) To assess the short-term impact of pandemic induced displacement on the migrants and their families mainly on livelihoods.
2) To document government initiatives in this regard to restore the livelihood and income of those people who are forced to revert back in villages.
3) To analyze the effectiveness of the current schemes and to provide strategies and recommendations on the way forward. Our ultimate aim is to raise the voices of these invisible citizens so that they can live a life of basic human dignity, not only in the short run but also in the long run.
4) To suggest some long-term solution for the migrants through generation of sustainable employment.

Livelihood Analysis of The Migrants who are Forcefully Reverting Back to Villages

The precarious situation of Covid-19 has brought to the fore the issue of humanitarian crisis like pandemic induced displacement. The unplanned lockdown in Indian economy has brought the conflict between lives and livelihoods of migrants in India. The Periodic Labour Force Survey (PLFS), Ministry of urban affairs and housing 2017, Census 2011 reveals the nature, magnitude and causes behind migration in India. From this data source India is estimated to have some 120 million rural to urban migrant labour. Interstate migrants are approx. 56 million and around 40 million of them were heading towards mega cities. Nine out of ten workers (93%) work in informal sector, eight of ten workers (79%) of migrants to the urban cities were based on daily wages. So, migrants are working in the urban areas without social protection and without ration card.

In India an already slowing economy before Covid- 19 exacerbated the crisis. This unplanned lockdown has devastated livelihoods at unprecedented levels, causing widespread job losses and hunger. There was a massive increase in unemployment and dramatic fall in earnings.

Key Findings

1) A reverse exodus of migrant laborers who are in a desperate situation without the basic necessities like food and shelter. They are forced to leave the cities as they are abandoned by their employers and contractors.
2) Seasonal migrants are the most vulnerable workforces of India.
3) The impact of lockdown has worsened the hunger situation and the noted fact is that India ranked 105th out of 117 countries in the Global Hunger Index in 2019 and it is further worsened.
4) Highest numbers of migrant workers are in the construction sector and they are basically daily wage earners. Construction sector absorbs a significant share of country's internal migrants. From NSS 64th Round we get the industry wise division of work for seasonal migrants and it reveals that 37% of seasonal migrants are involved in the construction sector.
5) Sudden announcement of lockdown has rendered workers unemployed across country and they are trapped in destination cities with inadequate resources to combat the situation and unable to pay rent and maintain social distances. They are exposed to hunger and improper hygiene and health issues. Displaces are in precarious situation with no money, no ration and someone has skipped the previous meal.
6) The welfare measures announced by the Central Government fail to address the concern for daily wage migrant laborers as they have no social protection and no ration card. They basically stuck in destination sites without employment and they are not entitled to get PDS in cities. Apart from this, many migrant workers are stranded in transit unable to reach their source. So, there is a consequent loss of livelihood and income for the seasonal migrants due to the immediate impact of Covid-19 lockdown.
7) These short-term findings have a long-term implication as the migrant workers are uncertain about their future ability to earn for their bare sustenance. This long-term impact should be taken into consideration by the policy makers.
8) Finally, rate of hunger and distress exceeding the rate of relief.

Policy Framework for India's Covid-19 Migration

From the key findings it is clear that vulnerabilities of the migrants in the context of pandemic induced displacement posed a major policy challenge for the government. This precarious situation exposed the unacceptable treatment of these migrants in Indian labor market even before the crisis starts and no Fair Migration Agenda is rooted in the policy making of the government. To explain the policy framework in this regard we have to extend the analysis in two phases:

1) Government Responses to Mitigate the Crisis
2) New Governance Challenges

Government Responses to Mitigate the Crisis

In this context various measures are taken by the government on income, ration and support.

Income Support

Due to lockdown the circular flow of product and income is snapped and supply chain is disrupted. The payment to the factors of production is disrupted and it has exposed glaring economic inequalities in the economy. In this context the role of the government is emphasized and even in the capitalist system the role of government is admitted. Direct financial assistance to the displaced people is required as migrants are excluded from social security measures provided by the state.

The scheme like Gareeb Kalyan Yojona aims to provide ration through public distribution system and financial assistance through Direct Benefit Transfer (DBT). Under this Yojona, there is a provision for fund transfer through DBT to women Jan Dahn Account holders and widow, elderly and disabled pensioners. In this context the importance of MGNREGA program has increased manifold and government has increased marginally the daily wage by Rs.20. Government has also issued a directive to transfer funds to the account of construction workers through DBT from the fund collected by Labour Welfare Board. This may play an important role for the livelihood issues of migrant workers. App based financial assistance for the migrants was introduced by the government of Bihar and Jharkhand.

Ration

The main problem for the migrants is that they have no ration card in their possession. Under Gareeb Kalyan Yojona, beneficiaries under National Food Security Act (NFSA) would be provided 5 kg of wheat or rice per person along 1 kg of pulses per household per month, in addition to the quantum that they are receiving prior to the crisis, free of cost. Finally, under Pradhan Mantri Ujjawala Yojona (PMUY) government will provide three LPG cylinders to the beneficiaries of PMUY between April and June 20. This will also be helpful for migrant workers who are displaced from the destination site. To overcome the non portability problem of ration cards basically for migrant workers, government's new initiative 'one nation one ration card' scheme will enable ration card holders to get their supply of food grains at any fair price shop across the country. The beneficiary verification through Aadhaar authentication will happen on the electronic point of sale (e-pos) devices.

Support Services

Considering the mass movement of migrant workers (in economic jargon which is known as reverse exodus) from destination sites, central government has instructed the states to immediately set up relief camps for migrant workers along with the spread of information about the location of the relief camp and National Disaster Response Force (NDRF) teams are deployed on routes taken by huge numbers of migrant workers. In nutshell these are different provision of relief measures taken by the government to tackle the problem of economic health (livelihoods) of the forced displaces in the context of pandemic. Expanding liquidity directly to micro small and medium enterprises (MSMEs) to protect employment has an important policy relevancy for migrants also.

New Governance Challenges

India's lockdown is an example of dominant global policy response to counter the Covid-19 pandemic. The philosophy behind the lockdown is life must be valued above commodities. To combat this menace there is a tradeoff between public health (lives) and economic health (livelihoods). In this regard public health gets the first priority. The cost of livelihood in this regard should be combated by the redistributive policy tilting in favour of alleviating the burden of the poor and the role of the government comes into play. To protect the most vulnerable of the population adequate transfer is required to tide over this difficult period. The aforementioned policies are taken by the government in this regard.

But any relief plan basically contains two aspects. One is distributive aspect and other is macroeconomic aspect. So, any evaluation of the relief plan is based on distributive and macro dimensions. By macro dimension we mean whether the relief plan helps to prop up the demand (fiscal measures) or manages the supply side issues (monetary measures). We are now switching from positive economics to normative economics. Various lists are potentially available for making targeted transfers like Jan Dhan Yojona (a financial inclusion program of the government), PDS, NREGA, BPL (below poverty line) etc which are earlier mentioned. But the problems in these targeted transfers are as follows:

- In BPL list there are significant errors of inclusion and major displaces do not have access to government ration.
- PDS and NREGA lists are imperfectly linked to bank accounts. Sometimes this correlation between targeted beneficiaries and bank account is not done properly and some states have failed to execute this comprehensive mapping between the beneficiaries of targeted program and bank account.
- Problems in proving the eligibility by way of the cumbersome Aadhaar-based biometric identification procedure. There should be flexibility in options of availing economic relief.
- Low rate of job card holders in the case of NREGA is the symbol of loss of trust in the system as there is a delay in payment release, low wages and prevalence of corruption. So, the issue of vulnerability remains for the migrant workers after the attachment of beneficiaries with the bank accounts.
- There are many unregistered labors who have no BOCW cards which means they would be ineligible for any benefit transfer.
- No macroeconomic policies to overcome the failure of supply chain mainly agricultural supply chain. As mandis are closed, it choked off the wholesale buying points in the chain. Freeing up of mandis and all markets for agricultural produce can somewhat

mitigate the livelihood problems for returning migrants from urban areas.

The relief measures proposed are quite inadequate. Government has allocated just a little below 1% of GDP to the relief plan. As there are some limitations in the cash assistance, government can augment its kind assistance in this regard. On May 12, government has announced its second relief plan which envisages an expenditure of round Rs.20 trillion. In this package, two schemes are worthwhile for displaces, one is the promotion of affordable rental housing complexes for migrant workers and urban poor and other is allocation in kind for non ration card holders and acknowledgement of the use of one nation one card. In this package government has increased the budget estimate of additional Rs.0.4tr to be allocated to NREGA employment over the earlier budget estimate to reduce the plight of the displaces.

The major pitfall of relief plan is that it is inadequate to the targeted beneficiaries in compare to the spending allocated by the developed country as a % of GDP. Even in per capita terms this fiscal stimulus is meager in terms of its population and vulnerability. In terms of relief plan 2, fiscal component is again small and very unclear. In this context it is clear that government basically moves towards easing of loans and liquidity which is basically a monetary measure in a situation when the country already is flooded with liquidity and uncertainty prevails in the context of investment.

Challenges Forward

Forget Covid, we will die of hunger first- media report shows the gravity of the problem for the displaces

From this pandemic induced displacement, it is clear that in cities, migrants are being reduced merely to labour power and excluded from all social safety mechanisms. The stimulus package announced by the government is to remedy only short-term worries and not to reduce vulnerabilities in long term. The future challenges for the government are:

- Government has to ensure universal access to public distribution system at this extraordinary time. This will take care of the social exclusion problem as social cost of the excluding the person who has the urgent need of PDS is greater than the social cost of including someone who could perhaps sustain without it.

- Government has to provide substantial fiscal stimulus as the ongoing demand crisis has further exacerbated due to pandemic. Mere injection of liquidity should not solve the problem. In this situation impotence of monetary policy by liquidity easing to banks is highly unlikely to be passed in terms of fresh loans to firms and this will highly trigger the role of fiscal policy. The distributive policy and fiscal stimuli go hand in hand. The government should provide money to the targeted group and this will return as an increase in aggregate demand.

Provision of shelter for migrant workers in urban areas, opening of community kitchens, price control for essential commodities, public hygiene education, extensive Covid 19 tests and free provision of health care services are further prescriptions for intervention of the government which can reduce the displacement problem temporarily.

- Parikshit Ghosh [4] advocates a social safety net in the form of a quasi-universal basic income, which should be paid over the next six months.
- Fund should be transferred to the states as collection from goods and service tax has declined. State spending will generate additional demand and state can better serve the targeted people as they have correct information on local conditions and can react in more flexible way than centre. This will solve the problem of informational failure in this regard.
- NREGA could expand its work guarantees well beyond 100 days per annum and NREGA should be complemented in cities also so that misery of displaces should be controlled off.
- As there are various reservations for cash transfer in JDY, the dependence on NREGA and PDS has been increased as the lists are much more transparent for cash transfer. Government has to stipulate the amount of cash transfer in this regard so that it can secure the basic needs of the displaces. There is some sort of reservations in giving cash transfer. Some economists opine that it is a transitory income and should not be used to increase aggregate demand. Someone has questioned regarding its funding. In this context it is true that though it is a transitory income it will help to increase aggregate demand as poor people has a higher C/Y ratio and they will try to smooth their

consumption first. In the context of funding government has the provision to borrow as private investment has been weak in this situation and there is a lesser chance of crowding out. Immediate relief measures are critical to the sense that economic activities will take at least few months before they can resume original capacity of production and displaces has to remain in their source places in this time interval.

It is imperative to move beyond the short-term worries of the livelihood problem of displaces just by providing immediate relief measures rather we should think of long-term development of the migrants so that we can resist the vulnerabilities of the displaces in the event Of Crisis.

Conclusion – Apprehension and Hope for Future

Jobless workers mull return to cities- The Telegraph, 2nd July, 2020.

Apprehension 1 is based on the deplorable conditions of the migrants who are returning to home from their destination. This crisis like this has shown as first hand just how critical and complex the coordination between the home states of migrants and the states where they work. Apprehension 2 is based on the decline in work provided through NREGA which is referred to as 'employer of the last resort'. According to the Centre for Monitoring Indian Economy (CMIE), rural unemployment rose from 8.49% in the end of March to 22.67% on April 29.

Apprehension 3 comes from the newspaper heading. Meager job opportunities and expectation of lower wages than what they had been accustomed to get, prompted many migrant workers to start making plans to go back to cities. So, the problem of no work and low pay at home.

There are two types of internal migrants. The first are those who are actually in need of work for sustenance while others are those who go to cities to augment their income. These latter types of migrants never going to return from the village now and there is a large scope to use them if government has prompted to its action of restoration of livelihood and asset creation in rural areas.

In this context government has to emphasize on the policies of livelihood support for long term after this type of reverse exodus.

The government should bring a road map of how they are going to employ such large number of migrants in the development work who are returning home. There is a huge loss in the demographic dividend if government fails to exploit this work force. A data base of all the returning migrants should be made and on the basis of that job should be provided to them. In this regard one has to be informed that there was no central registry of migrant workers despite the existence of Interstate migrant Workmen Act, 1979. In this context government should revamp the NREGA program in such a way so that number of days of guarantees of work per household should be increased and allocation of the work should also be increased. It is the time for government to use NREGA. It is true that it is difficult for them to establish their livelihood in the source region as they are the new comers in this region but government has the opportunity to revive the rural economy by using this work force. A nationwide large-scale employment guarantee program should be launched to combat the problem with urban wage employment programs. It is the high time for thinking the propaganda of using this surplus labor to build self reliant India and to build country's infrastructure and agricultural base that is returning to the root of the country. The crisis will go on until we initiate a major structural change in our economic policy. The role of Self-Help Group for using the collateral free credit as part of the program of PMGKY has a huge role to play. This will help to generate local demand and the problem of supply chain disruption can be addressed as locally available raw materials should be used in the production. It is true that government has ensured increasing liquidity through its stimulus package and now it should be balanced with large fiscal intervention. To relieve pressure on the rural banking system, panchayats could be used for cash transfer, as it was happened in Odisha. A mixture of cash payment and Food for work program that is kind payment in the case of NREGA workers can help to reduce the economic distress for the migrants. This will also help in proper utilization of the huge food reserves for Indian economy.

Sustainable Employment - Long Term Solution

To mitigate the crisis of the migrant workers there is the urgent need of the creation of a National Urban Employment Guarantee Program that strengthens small and medium sized towns in India by providing legal right to employment for urban residents. This will help to improve the quality of urban infrastructure and services, restoring urban commons and ecology, skilling youth and increasing the financial and human capacity of urban local bodies. This proposed program seeks to address the problem of migration to large cities from small and medium towns. The program should have a strong legal basis in the form of National Urban Employment Guarantee Act and should be demand driven and it provides statutory right to employment at specified wage rates. MGNREGA and this proposed program would go hand-in-hand. Earlier urban employment schemes were not demand driven and only a set of identified beneficiaries could avail their benefits. In the case of demand driven there is no need to identify the beneficiaries which is a complicated process in a country like India. It is important to improve the livelihoods and ecology of urban areas beyond India's major cities. This will also reduce the tendency of migration to large cities and can raise the living standard of the urban poor. Kerala has been running a program called the Ayyankali Urban Employment Guarantee Scheme (AUEGS) which guarantees 100 days of wage employment to an urban household for manual work. Apart from sustainable employment creation, government can issue "Covid bonds" and can finance the resulting debt through a progressive tax on the wealth of the richest 1% of the population. There is no burden of tax targeting in this case and this is a non inflationary mechanism for debt financing and a better method than deficit financing. This is one of the most sensible and humane prescription to raise resources to tackle the epidemic keeping fiscal deficit within control and government can use that fund for direct spending on the migrants who are forcibly displaced.

References

Arabinda Acharya and Nilachala Acharya. Covid -19: Can reverse Migration help Revive Rural Economy of Odisha? Opinions (May 7, 2020). URL: https://www.cbgaindia.org/blog/covid-19-can-reverse-migration-help-revive-rural- economy-odisha/

Kabir Agarwal. "COVID-19 Lockdown: In April, MGNREGA Work Crashed to Lowest in 7 Years". The Wire (May 1, 2020). URL: https://thewire.in/labor/covid-19-lockdown- mgnrega/

Abhinash Borah, Sabyasachi Das, Aparajita Dasgupta, Ashwini Deshpande, Kanika Mahajan, Bharat Ramaswami, Anuradha Saha, and Anisha Sharma. "Covid-19: Are We Ready for the Long Haul, Parts I and II". Ideas for India (Apr. 2, 2020). URL: https://www.ideasforindia.in/

topics/ macroeconomics/ covid-19-are-we-ready-for- thelong-haul-i.html.

Parikshit Ghosh. "We Need a Marshall Plan to Fight Covid-19". Ideas for India (Apr. 8, 2020). URL: https://www.ideasforindia.in/ topics/ macroeconomics/ we-need-a-marshallplan-to- fight-covid-19.html.

Dipa Sinha. "Food for All During Lockdown: State Governments Must Universalise PDS". The Wire (Apr. 20, 2020). URL: https://thewire.in/ rights/ covid-19-lockdown-foodsupply- pds.

Advaidh Nelakanttan R. "Why the COVID-19 migrant crisis should compel us to re-think urbanization?" (May 23, 2020) URL: https://www.cnbctv18.com/ views/ why-the-covid-19-migrant-crisis-should-compel-us-to-re-think-urbanisation-5987901.html.

Amit Basole, Mathew Idiculla, Rajendra Narayanan, Harini Nagendra and Seema Mundoli. "Strengthening towns through sustainable employment: A Job guarantee Programme for Urban India. State of working India 2019, Azim Premji University. URL:https://cse.azimpremjiuniversity.edu.in/ wpcontent/ uploads/ 2019/ 04/ SWI2019 _Urban_Job_Guarantee.pdf.

Debraj Ray and S. Subramamian. Covid-19: Is There a Reasonable Alternative to a Comprehensive Lockdown? Ideas for India, Mar. 28, 2020.

Debraj Ray and S. Subramanian. India's Lockdown: An Interim Report. NBER Working Paper No. 27282, May 20, 2020.

S. Subramanian. "Doing the Maths: Why India Should Introduce a Covid Wealth Tax on the Ultra Rich". Scroll (Apr. 16, 2020). URL: https://scroll.in/ article/ 959314/ doing-the-maths-why-india-should-introduce-a-covid-wealth-tax-on-theultra-rich.

Voices of the Invisible Citizens: A rapid Assessment on the impact of Covid-19 Lockdown on Internal Migrant Workers; Recommendations for the State, Industry and Philanthropies. Jan Sahas, New Delhi, April 2020.URL: https://ruralindiaonline.org/ library/ resource/ voices-of-the-invisible-citizens/.

Bogumil Terminski. Development-Induced Displacement and Resettlement: Social Problem and Human Rights Issue, University of Geneva Research Paper No. 9, posted: 9th June,2013.

Miachel M. Cernea. Understanding and preventing Impoverishment from Displacement- Reflections on the state of knowledge, Journal of Refugee studies Vol 8, No.3, 1995, Oxford University Press.

Maitreesh Ghatak, Sandip Mitra, Dilip Mookherjee and Anusha Nath. Land Acquisition and Compensation in Singur: What Really Happened? pdfs.semanticscholar.org, March 29, 2012.

9

The Problem of Immigration in Assam and Its Impact

Kanchan Devi
M.Phil. Research Scholar, Department of Economics,
Rajiv Gandhi University, Arunachal Pradesh, India

Abstract

Immigration is one of the most burning problems of Assam. It is going to more serious day by day as it creates threats not only on the identity of Assamese people but also in national security. However, immigration has its history in Assam. But now a day the migrants from the neighboring countries like Bangladesh and Pakistan and inter-regional migrants from West Bengal and Bihar are causing the biggest threat of losing the cultural and linguistic identity of the local people of Assam. This paper tries to figure out the various causes of the immigrants of people to Assam and its resultant impact on the political, economic, social, and environmental climate of Assam. The study is done with secondary sources of data. The findings of the study state that the environmental crisis caused by population explosion in Bangladesh and relatively greater opportunities like lifetime income, acquisition of land, and the Asset of Assam are the main causes of immigration of people in Assam. The study also found that the problem of rapid immigration in Assam leads to a multidimensional impact on Assam threatening national integrity and unity of Assam. The study will be helpful to those students, academicians, research scholars, and policymakers, the government official who is interested in the field of population economics, demography, and other social sciences.

Keywords: Immigration, Assam, People, Problem, Bangladesh.

Introduction

Thus immigration & migration is not at all a new phenomenon in the case of Assam. But since the beginning of 20th-century immigration in Assam is posing the severe problem in the socio-political condition and economy of Assam. Migration in terms of basic immigration has a dual impact on Assam; socio-cultural & environmental impact and the growth of population. Immigration negatively affects the native residents because at primarily they are unfriendly to the native people, the labor market of the place where migration occurs unduly stressed, the loads on social infrastructure got increased, the crime and conflict in migrating place increase and deterioration of the working and living condition of the native people occur (Irina Pavlona, 2015)[1].The census of 2011 estimates the total population of Assam as 26.6 million, which is very high in terms of its area. Again, raising the density of population from 286 per square kilometer in 1991 to 340 per square kilometer in 2001 depicts the picture of the rapid growth of the population in Assam. The growth in the population may occur due to two reasons_ one is due to natural increase and the other and most important is because immigration is a natural phenomenon and in Assam, it has its history. In the British period, laborers were imported from states like Bihar, Orrisa, Chotanagpur, madras, west province and other parts of India to work in the tea garden of Assam. Again, the agricultural farmers, which are setting on both the valleys of Assam, are largely coming from the districts of East Bangle. The environmental crisis caused by population growth in Bangladesh and greater economic opportunities is the factors for the high influx of population from Bangladesh to Assam (Hiranya Nath and Suresh Kr. Nath, 2011)[2]. The rapidly increasing trend of immigration creates a situation of vulnerability and threat among the indigenous people living in Assam. The excessive immigration in

1 Tsapenko, I. (2015). social effects of immigration. *Herald of the Russian Academy of Sciences*, 443-452.

2 Hiranya Nath and Suresh Kr. Nath. (2011). Illegal migration into Assam: Magnitude, causes, and economic consequences. *Sam Houston State University Department economics working paper.*

Assam has a multidimensional effect on Assam. In Assam, internal and illegal international migration of massive scale leads to existential fear and apprehension among its local people (Chandan Kumar Sharma, 2012)[3].The immigration of the people from Bangladesh to Assam leads to a threat like the situation of losing the identity of the people and the fear of being reduced to a minority within their homeland is the main concern of the native people (Nandita Saikia and William Joe, 2016)[4] .In this paper, I have analyzed the problem of immigration with its impact in Assam.

Area of the Study

The study area of this paper covers state Assam with special emphasis on some selected districts where the problem of immigration is more acute. Both internal migrants and international migrants are taken as immigrants in Assam in this study. Assam is a state from the northeastern part of India, which shares an international border with countries like Bhutan in the north and Bangladesh in the southeast. Assam is sharing its boundary with six neighboring states like Arunachal Pradesh in the northeast, Nagaland in East, Manipur in Southeast, Meghalaya, and Mizoram in South, Tripura in Southwest and West bangle in West. The whole northeast region is sharing a boundary with 1879 km with Bangladesh.

Objectives

The main objective of this paper is to focus on one of the important demographic problems of Assam i.e. the problem of immigration, to explore the reason for such influx of the population to Assam, and finally to analyze the impact of the immigration of the people in Assam.

Methodology of The Study

The descriptive paper is completed by studying extensive literatures and different facts and data are collected from various secondary sources including journals, periodicals, articles, and research papers.

3 Sarma, C. K. (n.d.). The immigration issue in Assam and conflicts around it.
4 Nandita Saikia and William Joe. (2016). *Cross Border migration in Assam during 1951-2011: Process, magnitude, and socio-economic consequences.* New Delhi: ICSSR.

Findings

From the study of Weiner (1973)[5], Goswami (1985)[6], Gogoi (1984) [7]and Hiranya Nath and Suresh Kr. Nath (2011) following reasons are found for continuous immigration in Assam from the pre-independence period:

1) The tea industry in Assam is the reason for immigration in Assam from the British period.
2) In older days in Assam, the high-ranking officials were imported from other parts of India to Assam.
3) From densely populated neighboring countries many migrant's influx in Assam for better economic opportunities.
4) Many migrants from east Bengal came to Assam to live near Brahmaputra valley for the extraction of economic interest.
5) To get rid of the land tenure system like the Zamindari system, immigrants came to Assam for permanent settlement.
6) The political system of India give chance the Hindu migrant came from East Bengal to settle in Assam.
7) The economic crisis of Bangladesh is the main reason behind the immigration of Muslim labor in districts like Goalpara, Kamrup, Darrang, and Nowgong.

The problem of immigration has a multidimensional impact on the economy, politics, and socio-cultural issues and along with the environment.

The Impact on Politics Includes Assam Agitation, the Problem of the Illegal Voter, and Terrorism

- **Assam Agitation:** The failure of the government in providing security to its citizen and the failure of democracy leads to the agitation of people against of government. At present ASSU is playing a significant role against CAA on behalf of the public.

5 Weiner, M. (1973). Assam and Migrants. *Demography of India, vol II.2.*

6 Goswami, H. (1985). *Population trends in the Brahmaputra valley (1881-1931).* New Delhi: Mittal publications.

7 Goswami, Atul, and Gogoi J. K. (1984). Migration and Demographic Transformation of Assam (1901-1971)i. In *N.E.Region: Problems and the prospect of Development, CRRID.* Chandigarh: n B.L.Abbi (Ed.).

- **Illegal Voters**: From the final draft of NRC in Assam it is seen that most of the illegal immigrants already got their name on the list. The majority of people who are immigrants claimed themselves as citizens of Assam making NRC useless in the identification process.
- **The Problem of Terrorism**: Government of India & Assam at many times seen as a failure in meeting the demands of some section of the society. So, deprived people often form a terrorist organization. One of such terrorist organization is Pakistan's ISI, which always supports Bangladesh against the militant movement in Assam.

Impact on the Economy Can Be Explained in Terms of Unemployment, Inflation, and Raising the Financing Burden on the Government

- **Rising Unemployment**: The problem of the rising unemployment rate is one of the major economic problems experienced by not only Assam but also by India due to excess population.
- **Poverty**: Poverty is also a macroeconomic problem faced by Assam & India originally and for immigration, the reduction of poverty is proved as a serious challenge for India.
- **Inflation**: In Assam, the growth of the population is mostly propelled by the immigration of the people and because of this excessive growth population there is excess demand in the economy rising price level.
- **Increase in Financial Burden on The Part of Government**: Due to growing population rate both the central and state government is facing a crisis in the management of the budget. Expenditure of the government is generally much higher than revenue in the budget of both sets of government.

Socio-Cultural Impact Because of the Immigration Can Be Best Explained by Following Headings: The Crisis of Identity, Community Tension

- **Crisis of Identity:** In Assam, the influx of immigration from different parts of within and from outside the country has been creating a situation of the threat of losing the identity by the indigenous people of Assam. The indigenous people in Assam have not only to share their land and other rights but also, they might lose their cultural and linguistic identity due to the immigration problem.

- **Community Tension**: At present not only in Assam but also in whole India the community tension is seen. Even the capital city Delhi is burning due to community tension, which arises due to the immigration problem endangering the peace and security of India & Assam.

Impacts of Immigration on the Environment can best be Explained by the Following Points: Environmental Degradation, the Problem of Poachers

- **Environmental Degradation**: Immigrants generally place in the area nearby forest, hills, rivers; which are all environmental goods. To get benefit from environmental good they destroy forests & hills, so several environmental problems like soil erosion, climate change, and flood are more severe year after year in Assam and some parts of India.
- **The Problem of Poacher**: The problem of poaching can also be attributed due to immigration, because which population grows at a rapid rate, and the government of Assam is unable to fulfill the needs of all people. Due to in availability of jobs some people use unfair means for livelihood to earn easy money. They even do not bother to destroy some valuable resources which are to be preserved. An example can be taken as the poaching of rhinoceros in the Kaziranga national park of Assam.

Discussion

In this period of the 20th century, Assam has experienced the highest growth of population among all the states of India. The data relating to the period of 1901 to 2001 reveals that the population of India during that period grew by 311% which is less than even half of the growth of the population of Assam, which is 710%. The large difference in the figure can well explain that massive immigration took place in Assam from densely populated neighboring country Bangladesh from 1991 to 2001. Due to various reasons that may be the environmental crisis in Bangladesh and the greater socio-economic opportunities in Assam, attracts a huge number of migrants to Assam. The following table shows the trend of population growth in the top immigrant prone district of Assam.

From the study it is found, following nine districts of Assam are facing a severe problem of immigration at present compared to other districts. The following table shows the decadal growth of the population of those districts in percent.

Table 1: Immigration Prone Districts with Decadal Growth of Population

Districts	Decadal Growth Rate In 2001 (In %/)	Decadal Growth In 2011 (In %)
Dhubri	22.97	24.40
Goalpara	23.03	22.74
Barpeta	19.62	21.40
Morgan	21.35	23.39
Nagaon	22.26	22.09
Dhemaji	19.45	20.30
Cacher	18.89	20.17
Karimganj	21.87	20.74
Hailakandi	20.89	21.44

Source: Census of India, 2001 & 2011.

From the table, it is seen that there is decadal growth in all most all the migrant prone districts of Assam from 2001 to 2011. This growth of population during the decade is not only due to natural increase but also because of the immigration of the people to Assam from the neighboring states and countries to find better opportunities for livelihood. In context to international migrants especially from three countries viz. Pakistan, Bangladesh, and Nepal people migrate to Assam for a different reason. In 1991, total immigrants from Assam from Bangladesh records 2, 88,109, from Nepal 22,433 and Pakistan 15,110(Census of India, 2011). Due to different economic, social, political, and environmental crises, people from those countries came to Assam for better livelihood. In the case of inter district, most of the people came to Assam from the states like Orissa, West Bengal, Bihar, Rajasthan, etc. for trade, business, and commercial purposes.

But the high inflow of people to Assam creates undesirable changes in the socio-cultural, economic, political, and environmental well-being of the local ethnic people of Assam. The most important is it creates a challenge for unity and integrity of Assam, as indigenous people of Assam become feels insecure of their right and liberty. The fear of losing their socio-cultural identity threat of being a linguistic minority in their place is a matter of concern for them. The migrant people both legally and illegally are living in the area of Assam where the source of livelihood is relatively greater. So, they are living nearby rivers, forests, hill, and other such areas where they can start the business. This leads to the destruction of forest areas; pollution of rivers and the problem of poacher arise. Assam has reached already a limit to population growth, poverty and unemployment is a serious challenge for Assam. In such a situation influx of people from other parts leads to the unavailability of the government facilities for the local people as they have to share with immigrants. Though illegal immigrants do not pay taxes but enjoy all the services made for the general people of Assam. These lead to a burden on the part of the government as it has to serve all the people of Assam.

There is no denying fact that immigration has a strong contribution to creating the modern sector in Assam from the colonial period. The cheap imported labor helped in the plantation industry and development of other sector i.e. agriculture and industry. But in the 1980s the feelings of nationalism and Assamese identity emerged among the local people of Assam. Accordingly, different civil wars and movements happened in that decade. Finally, Assam accord was signed between the government of India and All Assam Student Union (ASSU), the super header of 1979-85 movement and Assam Gana Parishad (AGP), regional political party setting up of 24th March 1971 as the cut off the year to identify the illegal immigrants came in Assam. But the identification of illegal immigrants had not proceeded from that year itself and it required much time. But in course of time different political parties framed new laws and amended citizenship act in time to time with their intension. Due to such flexible laws, year after year immigrants from different parts of India and outside India are coming to Assam and a question on the identity of original people of Assam arises.

At present also Assam is burning due to recently brought citizenship amendment act (CAA) of 2019, amending the existing one.

CAA i.e. Citizenship Amendment Act is all about giving citizenship to six religious minorities mainly Hindu, Sikh, Buddhist, Jain, Persi, and Christian from the neighboring Muslim majority countries of Bangladesh, Pakistan, and Afghanistan, who are religiously persecuted and who came to India by December 31, 2014. Because of this act whole India along with Assam is suffering from clash and conflicting situations and tension between Hindu and Muslim religious people. According to many prominent people, the act is unconstitutional as it violates the basic structure of the Indian constitution i.e. secular, equality, and democracy principle is affected by this act.

Conclusion

The problem of migration is posing a serious threat not only in Assam but also in other parts of the northeast region except the protected areas under NEFA (North-East Frontier Area). The number of illegal Bangladeshi immigrant's rates is much more in Assam than any other state in India. If not properly checked; then it may swap Assamese identity from the northeastern part of India. But in the case of Assam, the role of government is critical as it is making such provision to give citizenship to some extra people violating the Assam accord in the name of religious persecution for vote bank. So, every educated & responsible citizen must make aware about the illegal immigrant problem and its consequences in Assam. This is urgently required not only to protect the identity of indigenous people but also to maintain national security. To reduce the entry of illegal immigrants into Assam border fencing should be strictly implemented and the areas, which are generally more immigrant prone, in such area security should be stricter and to identify people provision of valid photo identification with strict laws should be enforced. With this, we can conclude that the problem of immigration into Assam should be treated as a national problem than a regional issue.

References

Assam, G. o. (8 November 1998). *Illegal immigration into Assam, submitted to the president of India.*

Bandyopadhyay.S. (1999). Migration in the Northeastern region of India during 1991-97: Size trends, reasons, and impact. *vol.28.*

D, D. J. (n.d.). Socio-economic and political consequences of illegal migration into Assam from Bangladesh. *Journal of tourism and hospitality.*

Goswami, Atul, and Gogoi J. K. (1984). Migration and Demographic Transformation of Assam (1901-1971)i. In *N.E.Region: Problems and the prospect of Development, CRRID.* Chandigarh: n B.L.Abbi (Ed.).

Goswami, H. (1985). *Population trends in the Brahmaputra valley (1881-1931).* New Delhi: Mittal publications.

Gupta, D. T. (2016, 01 23). Immigration & its impact on Assam.

Hiranya Nath and Suresh Kr. Nath. (2011). Illegal migration into Assam: Magnitude, causes, and economic consequences. *Sam Houston State University Department economics working paper.*

Nandita Saikia and William Joe. (2016). *Cross Border migration in Assam during 1951-2011: Process, magnitude, and socio-economic consequences.* New Delhi: ICSSR.

Saikia, A. A. (SEPTEMBER, 2O017). Illegal Bangladeshi: Conflict ain political context. *Journal Humanities and social science, VOL.22.*

Sarma, C. K. (n.d.). The immigration issue in Assam and conflicts around it.

Sarma, M. (March 2015). A study of migration from Bangladesh to Assam, India & its impact.

Sharma, C. K. (2012). The immigration issue in Assam. *Asian ethnicity,* 1-23.

Tsapenko, I. (2015). social effects of immigration. *Herald of the Russian Academy of Sciences,* 443- 452.

Weiner, M. (1973). Assam and Migrants. *Demography of India, vol II.2.*

10

Environmental Degradation and its Impact on People of Bhubaneswar

Subhasmita Parida[1] & Dr. Sudhir Kumar Kisku[2]

[1]Ph.D. Research Scholar, P.G. Department of Geography,
Utkal University, Vhani Vihar, Odisha, India.
[2] Assistant Professor, Department of Chemistry,
M.P.C Autonomous College, Baripada, Odisha, India.

Abstract

The environmental degradation or crisis caused due to environmental and ecological changes is mostly by the result of developmental processes of the 'economic and technological man' of the present century. The most striking reason of the environmental degradation and hence global environmental crisis is the fast deteriorating relationship between man and environment because of rapid rate of exploitation of natural resources, technological development and industrial expansion. Increasing urbanization also results in marked reduction in arable land and therefore in food production. Thus, the rapid rate of increase of world population, ever increasing concentration of population in the urban centres, rapidly increasing gap between the populations of the developed and developing regions, increase proportion urban-rural population etc. are the basic and root causes of all sorts of environmental degradation and ecological imbalance because increase in population means

[1]Corresponding Author

more exploitation of natural resources which means inauguration of new environmental problems. Overpopulation also results in economic poverty because all the resources are utilized to meet the bare necessity of the people. This results in unbalanced exploitation of natural resources and subsequent environmental degradation. Industrial expansion, modernization of agriculture, urban growth, expansion of transportation network, scientific and technological development consequent upon alarming rate of population growth through decades is the major cause of environmental degradation. Since last few years the environment of Bhubaneswar city has been degraded due to the above mentioning man made activities due to alarming rate of population growth.

Keywords: Environment Degradation, Population Growth, Slum, Satellite Town.

Introduction

Environment can be defined as the physical surrounding of man/ woman of which he/she is a part and on which he/she is dependent for his/her activities like physiological functioning, production and consumption. His physical environment stretches from air, water and land to natural resources like energy carriers, soil and plants, animals and ecosystems. Environmental degradation is the disintegration of the earth or deterioration of the environment through consumption of assets, like, air, water and soil. The rate of environmental change and resultant environmental degradation caused by human activities has been so fast and widespread that R.F. Dasmann (1976) has remarked that 'the human race is like an ape with a hand-grenade. Nobody can say when he will pull the pin of the grenade and the whole world will be destroyed.

Population Growth and Environmental Degradation in Bhubaneswar

The concentration of populations in urban centres and in the developing regions is growing like mushroom both in the past and present century. Similarly, ever increasing concentration of population in the urban centres causes slums and environmental degradation and pollution. Increasing urbanization also results in marked reduction in arable land and therefore in food production. Increasing urbanization also results in the siphoning of resources from the rural areas to the urban areas which causes economic imbalance. This resultant economic imbalance causes economic and social pollution. Bhubaneswar city was originally planned for 40,000 people in1948 with an area of 16.48 km² with a density of 10 to 12 families per acre, has grown to about 148 km² with a population of 837,737 as per the 2011 census.

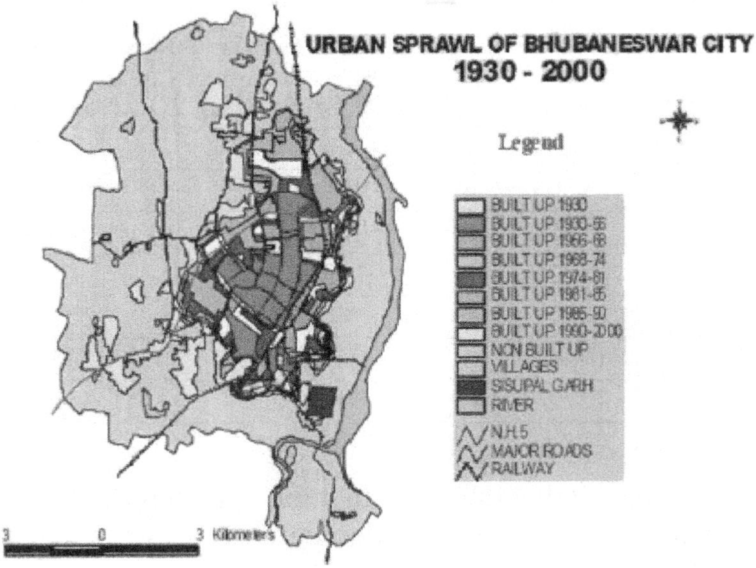

Figure 1: Urban Sprawl of Bhubaneswar City from 1930

Broadly, the environmental degradation of Bhubaneswar city due to rapid increase in the human population may be ascribed to:

1) Accelerated pace of educational and technological development
2) Ambitious developmental projects aimed at fast economic development
3) Firstly expanding industries, sprawling urban growth and agricultural development
4) Philosophical and religious outlook of the society
5) Cruel behaviour of man with the natural environment
6) Ignorance and lack of environmental perception and lack of public awareness towards environmental problems
7) Poverty
8) Affluence and richness

9) Unscientific and illogical exploitation and utilization of natural resources etc.

Trend of Population Growth

The entire municipal area of the city is divided into 47 administrative divisions called 'Wards'.

Table 1: Population Details and Growth Trends of Bhubaneswar City

Census year	Population	Decadal Growth (%)	Area (sq. Km)
1951	16512	-	25.90
1961	38211	131.41	50.20
1971	105491	176.07	65.03
1981	219211	107.80	92.91
1991	411542	87.74	124.74
2001	648032	57.46	135.00
2011	837737	45.90	151.96

Source: Census of India, 2011.

From the above table it is clear that the city registered the highest increase of about 131% during the decade 1951-61 owing to the sifting of the capital city of the state from Cuttack to Bhubaneswar. The city continued to grow and registered of 176.07% during the decade 1961-71. This was one of the highest growth rate experienced by any other capital city in the country and was also higher than any of the cities in Odisha during the decade.

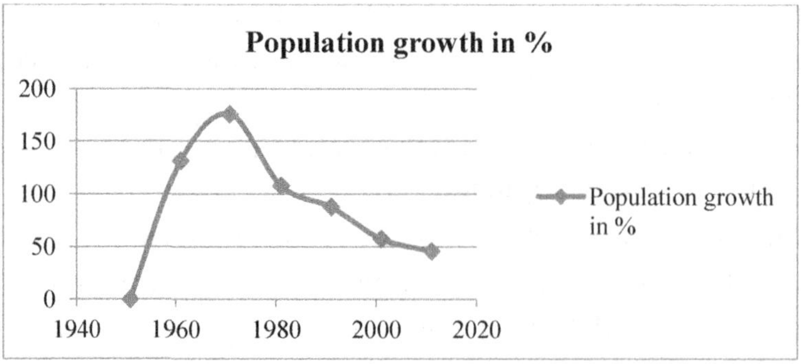

Figure 2: Population Growth Trend in Bhubaneswar City

Increasing Population Density

The population density of the city works out to about 5520 persons per sq. Km inside the municipal area of 151.96 sq. Km as per the 2011 census. The decadal growth of increasing population density of the city is shown in the figure below:

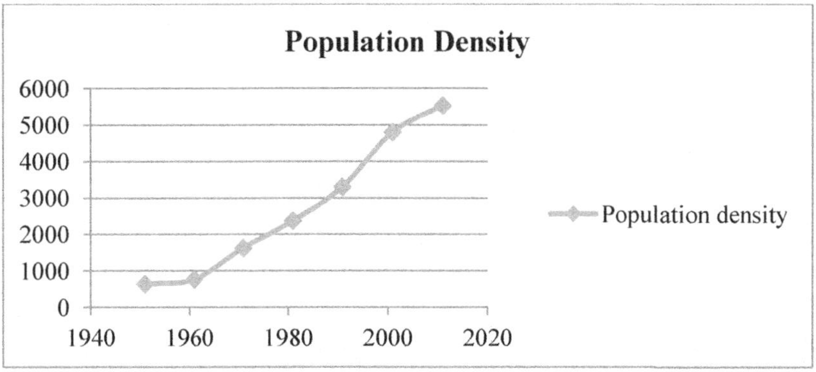

Figure 3: Growth in Population Density of Bhubaneswar City

It may be noted that the gross density has been increasing from 638 persons per sq. Km in 1951 to 5520 persons per sq. Km in 2011 as per the respective census figures.

Based on the ward wise distribution of population, the Bhubaneswar Municipality Corporation (BMC) area is divided into 5 zones:

Table 2: Ward-Wise Distribution of Population Density in Bhubaneswar City

Density	Ward No.
Very dense areas (20000-25000 Persons/ sq.km)	8,15,28
Dense areas (15000-20000 Persons/ sq.km)	5,9,11,23
Moderately dense areas (10000-15000 Persons/ sq.km)	10,12,16,21,22,24
Medium dense area (5000-10000 Persons/ sq.km)	4,6,7,13,14,25,26,27
Low dense areas (0-5000 Persons/ sq.km)	1,2,3,17,18,19,20,29,30

Source: BMC Profile, 2014.

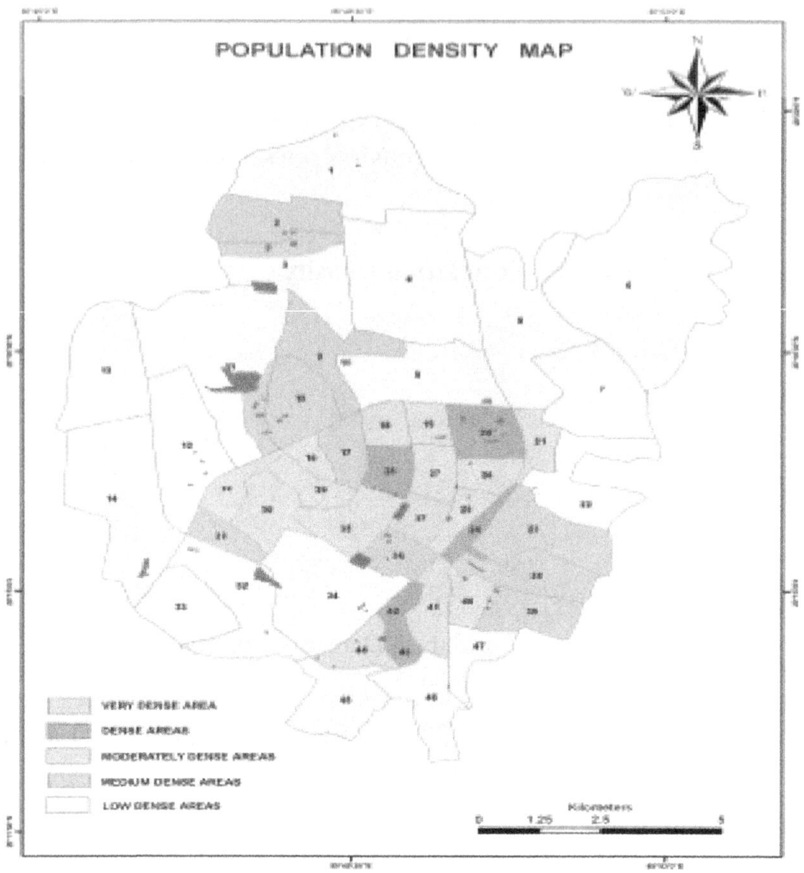

Source: GIS based map from Bhubaneswar toposheet

Figure 4: Population Density Map of Bhubaneswar

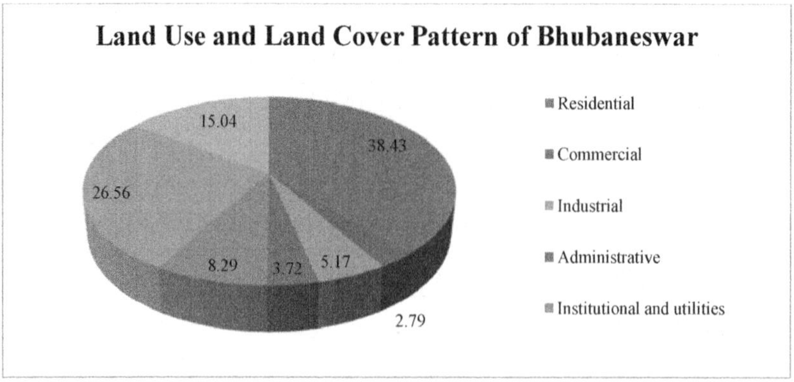

Figure 5: Land Use & Land Cover Map of Bhubaneswar

Industrial Activities

Bhubaneswar enjoys the status of being an important place in India in the production of various handicrafts, stone inlay and carpet craft. Apart from the cottage industry, there are 4 industrial estates are in Bhubaneswar city namely Rasulgarh, Mancheswar, Chandaka and Bhagabanpur. The total numbers of industries under operation in Bhubaneswar city are given in the table below:

Table 3: No. of Industries Under Operation in Bhubaneswar City

Name of industrial estate	Total no. of industries under operation			
	Large	Medium	Small	Total
Rasulgarh	1	1	4	6
Mancheswar		22	23	45
Chandaka	2	5	3	10
Bhagabanpur		1	8	1
Others		5	13	18
Total	3	34	51	88

Source: BMC Profile PDF 2015.

Profile of Slum Population

About one-third of the city's population is estimated to be living in slums and the slum pockets are distributed throughout the city. There was a tremendous growth (about 78%) in the slum population during the decade 1991-2001. This is mainly due to the vast devastation caused in the city by the super cyclone 1999, which has led to huge migration from the rural hinterland and other parts of the state as well as outside in search of employment particularly in construction sector.

Slums in the city are devoid of services, prone to all types of natural hazards, have poor living conditions and hence can pose severe environmental problems.There are about 377 major and minor slums located in and around Bhubaneswar city comprising total population of more than 2 lacks. Some of the major slums of Bhubaneswar are:

Salia sahi	Pathara bandha, Vanivihar
Bhoi sahi, Unit-9	Telugu basti, Saheed Nagar
Dumuduma	Science Park basti
Mundia sahi	Sikharachandi

Causes of Environmental Degradation

1. Improper Management of Solid Wastes

We all produce unwanted by-products and residues in nearly everything we do. Waste represents a valuable resource as ground cover to reduce erosion and as fertilizer to nourish new crops, but they also constitute the single largest source of nonpoint air and water pollution in the city. Improper disposal practices of solid wastes in the city are now resulting in serious and widespread pollution.

2. Municipal Solid Waste Generation and Storage

Bhubaneswar city generates approximately 466MT of municipal solid waste everyday comprising both bio-degradable and bio-non-degradable components. The total present generation of waste works out to a daily per capita of 400 grams, which is significantly higher than the normative standard of 250 grams per capita day for cities of similar scale. The major source of municipal solid waste are domestic, fruits and vegetable markets, hotels and restaurants, fish/meat shops and markets, street sweepings, hospitals, office institutions etc. Source wise reconciliation of solid wastes of the city is furnished in the table below:

Table 4: General Composition of Municipal Solid Waste Generated in Bhubaneswar

Source	Source-wise Break-up of waste generation in Grams			Percentage
	Bio-degradable	Non-biodegradable	Total	
Domestic	181.23	36.25	217.47	72.96
Hotel and Restaurants	16.80	2.23	19.03	6.38
Commercial	12.04	14.58	26.62	8.93
Institutions	9.94	11.15	21.09	7.07
Temples	5.07	0.25	5.32	1.79
Marriage Halls	7.29	1.26	8.55	2.87
Total	232.36	65.72	298.07	100.00

Source: "About Bhubaneswar". Bhubaneswar Municipal Corporation.

The table given below provides details on physical characteristic and concentration of heavy metals in the waste generated in Bhubaneswar city.

Table 5: Physical Composition of Municipal Solid Wastes Generated in Bhubaneswar

Components	Physical components of waste generation (Grams)		Percentage composition	
	Dry season	Wet season	Dry season	Wet season
Fermentable	346	234	34.67	23.28
Paper	65	94	6.51	9.35
Textile	72	87	7.21	8.66
Plastics	64	103	6.41	10.25
Combustible	211	121	21.14	12.04
Glass	38	103	3.81	10.25
Metals	33	97	3.31	9.65
Incombustible	113	129	11.32	12.84
Miscellaneous	56	37	5.61	3.68
Total	998	1005	100.00	100.00

Source: http://bmc.gov.in/

Anaerobic decomposition of solid wastes occurs inside the dumping yards which results in the emission of CH_4, CO^2 and other trace gases. These gases contribute to global warming and are one of the main reasons of high temperature in summer in Bhubaneswar.It is observed that, a major part of the above waste is collected and transported from each ward by the contingents of Bhubaneswar Municipality Corporation (BMC) throughout sourcing 14 out of 47 wards. Such partly segregated solid waste is presently transported to the nearest temporary solid waste disposal sites within the city.

3. Temporary Waste Storage Points

The BMC has provided 381 cylindrical bottomless concrete bins and these are located in the 47 wards. Out of 381 waste bins with a collective volumetric capacity of 90 cubic meters, 230 were observed to be damaged which signifies 60% de-rating of the collection infrastructure.

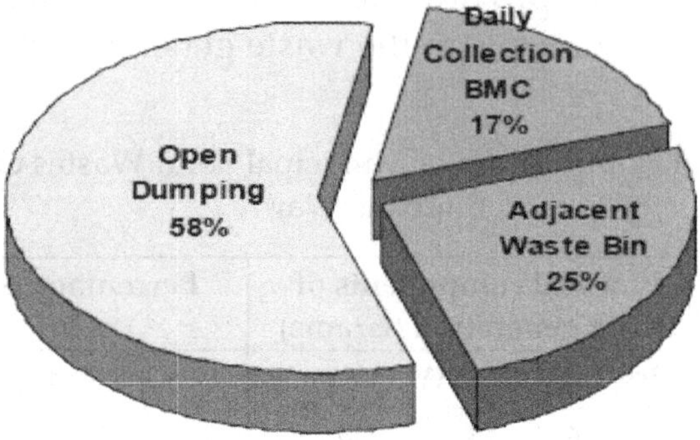

Figure 6: Means of Solid Waste Disposal in Bhubaneswar City

4. Poor Drainage System

The important rivers of Bhubaneswar city are Mahanadi, Daya and Kuakhai. These rivers supply around 80% of the total daily demand of potable water. There are 10 major natural drains running west to east of the city. Some these drains join the Ganguanallah which finally meets the river Daya. The entire Bhubaneswar city has not been covered with storm water drains. The drain length is about 363km though there is a requirement of 1000 km length of drain. Bhubaneswar is a complete hilly terrain, it is expected that within few minutes of rain, water should pass through the drains which ultimately will reach Ganguanallah. However such thing does not happens due to the following reasons:

- Encroachment of natural drains by construction.
- Dumping of waste material by the people making the storm water drains a dumping yard which results in blocking and chocking of drains.
- Lack of regular, maintenance of the drains i.e. desilting, emptying/clearing of solid wastes from the drains, repair of drains etc.

These are all the reasons of artificial flood in Bhubaneswar city. However, the rivers Daya, Mahanadi, Daya and Kuakhai are the major sources of water supply to the city that accounts to 182MLD (Million of litter per day). All the rivers and drains of Bhubaneswar city also carry storm water and treated sewage. Rivers Kuakhai and Ganguanallah receive about 107.25 MLD of wastewater of which 47.6 MLD is from domestic sources, 29.3 MLD is from industrial sources and 30.35 MLD is from mixed sources.

The entire drainage system of the Bhubaneswar city has shown in the figure below:

Figure 7: Drainage Map of Bhubaneswar City

5. Sewerage System

BMC is only partially covered with an underground sewage collection and conveyance system and as a result majority of sewage flows through open drains. River Daya, ultimately receives these huge pollution loads and is functioning like a 'major sewer'. Majority of the existing systems for sewage collection and treatment are not functioning well.The high rate of urbanization has led to water logging situation in some areas as infiltration to the sub soils reduced with increase in built up area.

It has been reported that about 80% of the existing sewer is inadequate or damaged and needs replacement or repair. About 20,000 manholes are provided out of which around 80% is damaged and needs replacement or repair work. Discharge of sewage into open drains at IRC village. The drain **No.1** opens into the river Kuakhai and the remaining

9 open drains (Nos.2-10) to Gangua nallah which in turn transfer the volume and load to River Daya.

6. Poor Sanitation System

As far as sanitation of Bhubaneswar city is concerned, it is pretty poor than other urban bodies of India. If we look at the slum profile of Bhubaneswar, then we find that more that 50% slums do not have access to both individual as well as community toilets. This clearly underscores the fact that almost 50% of the slum population and at least 40% of the city population does not have access to toilets and defecate in the open.

7. Deforestation/Cutting of Trees

Bharatpur Reserve Forest, spread over 1,200 acres, is located in the north-west part of the city. This is popularly known for its Chandaka Elephant Sanctuary and Nandankanan Zoo. There are also 3 major parks consstructed in the city. They are the Indra Gandhi Park spread over an area of 10 acres, Biju Patnaik Park and Gandhi Park. Other smaller parks are the Nalco Square, Nicco Park, Saheed Nagar, JK Park, Dr. Mugherjee Park and VSS Nagar.

Deforestation of Chandaka forest by super cyclone and also by local people for their livelihood contributed to the degradation of the environment. Cutting of trees /deforestation degrades environment in the following manner:

Trees absorb pollutants, NO_2, SO_2, O_3, CO through the pores of leaf surfaces.

Through their leaves and branches small particulate matter such as pollen, dust, smoke, and ash are trapped and filtered.

For Photosynthesis, trees absorb CO2 from atmosphere, oxygen is the by product of photo synthesis which is released by tree into the atmosphere. The roots of the trees increase soil permeability which results in reduced surface run off of water from storms, reduces soil erosion and increase ground water recharge.The air temperature also lowers through shade of the trees.

8. Weak Public Transportation System

The public city transportation system is very week in Bhubaneswar. Peoples use their own vehicles mostly two wheeler and depend on auto rickshaw to travel to different parts of the city. Most of the accidents inside Bhubaneswar is not only due to the poor driving of auto rickshaw drivers but also week repair , maintenance and high volume of around 18000 of auto rickshaws.

Impact of Environmental Degradation on People

1. Excessive Heat in Summer

Bhubaneswar has a tropical savanna climate, designated *Aw* under the Köppen climate classification. The annual mean temperature is 27.4 °C (81.3 °F); monthly mean temperatures are 22-32 °C (72-90 °F. Summers (March to June) are hot and humid, with temperatures in the low 30s C; during dry spells, maximum temperatures often exceed 40 °C (104 °F) in May and June. Winter lasts for only about ten weeks, with seasonal lows dipping to 15–18 °C (59–64 °F) in December and January. May is the hottest month, when daily temperatures range from 27-38 °C (81–100 °F). January, the coldest month, has temperatures varying from 15–28 °C (59–82 °F). The highest recorded temperature is 45 °C (113.0 °F), and the lowest is 12 °C (54 °F).

The average monthly temperature data (from January to August) of Bhubaneswar city is given in the table below:

Table 6: Temperature Data of Bhubaneswar City for The Year 2019 (January To August)

Month	Jan	Feb	Mar	Apr	May	Jun	Jul	Aug
Record high °C	35.8	42.7	42.0	45.0	46.5	46.3	41.3	37.4
Average high °C	28.7	31.4	34.9	36.9	37.2	35.3	32.2	31.6
Daily mean °C	22.2	25.1	28.6	30.9	31.7	30.7	28.7	28.4
Average low °C	15.6	18.7	22.2	25.0	26.2	26.1	25.2	25.1
Record low °C	8.6	9.6	14.6	17.0	15.0	16.8	20.0	18.4

Source: Meteorology department of Bhubaneswar

But in last 4-5 years, it has been found to be excessive heat in summer both during day and night. The temperature in summer is ranging in most of the days around 40^0C-45^0C which not only is affecting the health and comfort of the people but also is increasing the energy consumption by excessive use of air conditioner, refrigerators etc.

The main reasons of excessive heat in summer in Bhubaneswar city could be underlined as:

- Frequent loss and encroachment of wetlands and water bodies.
- High rate of air pollution due to increase in the use of vehicles.
- Large scale deforestation due to construction of roads and apartments.

2. Scarcity of Drinking Water in Summer and Artificial Flood in Rainy Season

Rains brought by the Bay of Bengal branch of the south west summer monsoon lash Bhubaneswar between June and September, supplying it with most of its annual rainfall of 1,542 mm (61 in). The highest monthly rainfall total, 330 mm (13 in), occurs in August.

The city receives around 1500 mm of rainfall a year, but the number of rainy days has been decreased from 70-82 days to 60 days. Most of the rainfall is due to cyclone than due to monsoon.

The opposite things happen in rainy season (even in some other seasons also) if high intensity rainfall occurs for some time flood like situations /artificial flood is seen in different parts of the city like Shastri nagar, Paika nagar, Ravi talkies area etc.

Even in month of October artificial flood has been seen in different part of Bhubaneswar. The example is 22nd October 2011 artificial flood in Acharya Vihar . Water get stagnated and does not find path to pass. This type flooding creates traffic jam, traffic diversion , accidents , water bodies gets polluted.

3. Water Pollution

The sources of water for Bhubaneswar are both surface and ground water. The surface sources are Kuakhai, Daya and Mahanadi River along with some lakes and ponds in Bhubaneswar. The ground water sources are open and tube well. The total water supply in the city amount to 260.78 Million litres daily (MLD) 18 .82% i.e.49.08MLD of water supply in Bhubaneswar is from 176 dug well and bore wells, the water loss is 40% due to leakage.

From the environmental profile of Bhubaneswar, it is found that:

- The characteristics of surface water as per the classification of central pollution control board fall in category of class - C which demands conventional treatment and disinfection.
- The lake and ponds water are also unfit for bathing purposes and rituals as it fails to satisfy the status- B category.
- The results of ground water quality analysis state that though the general characteristics of water conform to moderate quality, the coli form count in the case of open wells was found to exceed the normative level in some locations. In case of tube well, the low PH and presence of iron raises concern.
- The river Daya appears to be highly polluted due to discharge from Ganguanallah. The presence of faecal –coli form in river water indicates the presence of domestic sewage including human excreta.

4. Air Pollution

The different sources of air pollution in Bhubaneswar are stone crushers/crusher units, small scale industries, motor vehicles etc. It has been found that high to critical levels of SPM in most of the locations, SO2 within limits and NOx on moderate levels. However, any visible/ significant effect of Air pollution has not been felt or reported anywhere.

Table 7: Total Emission Load Per Day

S. No.	Sectors	Pollution Load					
		SPM	SO_2	NO_X	HC	CO	Total
1.	Domestic	0.61	0.21	0.37	0.06	0.07	1.32
2.	Industries	3.75	0.19	0.16	0.01	0.02	4.13
3.	Vehicles	0.41	0.35	2.51	0.62	4.75	10.64
4.	DG Sets	0.03	0.15	0.1	0.05	0.06	0.39
	Sub-total (1+4)	4.8	0.9	3.14	2.74	4.9	16.48

Source: Air pollution control Board Bhubaneswar

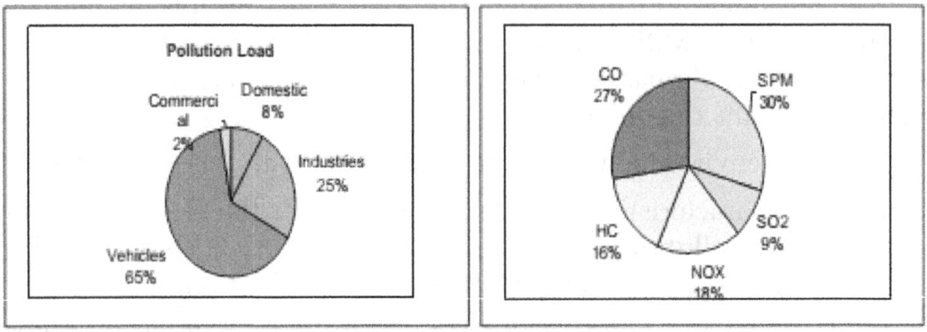

Figure 8: Total Emission from Different Sectors and Emission of Different Gases

Possible Reduction Measures

1. Increase in Vegetation Cover

Sufficient number of trees and other plants should be planted along the roads, along the drains, inside office building, inside residential buildings, in College and School campuses etc. People should be encouraged to establish horticulture garden around their buildings. Creation of new parks in each wards and maintenance of old parks should be done. Further destruction Chandaka forest should be prevented.

2. Introduction of Vibrant Public Transportation System

Mass Transportation system should be introduced which will reduce the fuel consumption and traffic congestion. The number of auto rickshaw should also be reduced. Old vehicles should be checked regularly. Frequent training to auto rickshaw drivers regarding traffic rules and driving should be organized. People should be encouraged to use alternative sources of energy such as solar, wind, bio-gas and bio-mass power etc. Dependencies on fossil fuels should be reduced. People should be encouraged to use bicycles, and bicycle tracks should be constructed.

Figure 9: Map Showing Existing Forest Cover and Developed Parks in Bhubaneswar

3. Proper Management of Municipal Solid Wastes

The collection efficiency of municipal solid wastes should be increased by engaging a greater number of persons, vehicles and equipments. There should not be any dumping yard inside the city. The waste should be disposed properly in sanitary landfill, constructed outside the city which should also be utilized to generate energy.

4. Construction and Maintenance of Efficient Drainage and Sewerage System

5. Restoration /Preservation of Wet lands and Water bodies

Wet lands and water bodies like lakes, ponds should not be encroached and they should be protected properly.

6. Improvement of Living Conditions in Slum Areas

7. Construction of Satellite Cities

To reduce the population density satellite city should be constructed with all modern facilities. More number of large markets, shops should be made in different places of Bhubaneswar to avoid traffic congestions in some particular areas. Area of Bhubaneswar Municipal Corporation (BMC) should be increased and work should progress speedily. Planning should be done such that housing, commercial areas, and offices should be located close to each other.

8. Encouragement for Rain Water Harvesting and Artificial Recharge of Ground Water

As far as possible ground water recharge should be done by both Government and public. Roof water harvesting should be encouraged to increase the level of ground water table and increase in ground water reserve.

9. Construction of Green Roofs and Cool Pavements

Peoples should be advised to construct green roof or roof top garden. A roof top garden is a vegetative layer grown on a roof top. Green roof provide shade and remove heat from the air through evapotranspiration reducing the temperature of roof surface and surrounding air. Cool pavements should be constructed by using pavement materials that reflect maximum solar energy, increases water absorption and remains cooler than the conventional pavements.

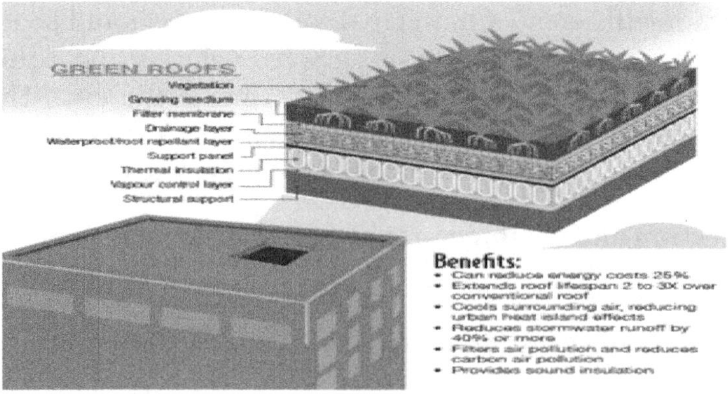

Figure 10: Diagram Shows How to Construct Green Roofs

10. Creation of Awareness and Strict Enforcement of Environmental Laws and Government Regulations

Conclusion

The degradation of environment of Bhubaneswar has not so far being alarming like metro cities of India but still it is a matter of concern particularly the heat in summer and unusual flood in the rainy season. The effects and causes of environmental degradation of Bhubaneswar city has been discussed in this paper. Some remedial measures have also been suggested. Government of Odisha has started to take some initiative though in late , to control the environmental degradation like plantation of trees in different wards, introduction of City bus service under JNNURM schemes, construction fly over in some important crossings, widening of different important crowded stretches, ban on entry of heavy vehicles during rush time, adding more number of revenue villages under BMC for construction of mass housing, construction of vending zones by removing the encroachment of roads by small vendors and cabin, distilling of storm water drains etc. It is the duty of not only the government but all citizens to be involved in different activities which will prevent the degradation of the environment of Bhubaneswar city and improve its quality.

> "We shall soon experience an environmental collapse unless a grass-root movement makes it imperative that public and the scientific establishment gives high priorities to the study and control of the forces that are rapidly making the earth a place unfit for human life."

References

Abraham, Priya (14 April 2008). "Glorious temple town growing". The Telegraph. Retrieved 24 December 2012.

"About Bhubaneswar". Bhubaneswar Municipal Corporation. Retrieved 26 December,2012.

"BDA Snapshot" (PDF). Bhubaneswar Development Authority. Retrieved 26 December, 2012.

"Bhubaneshwar Climate Normals 1971–1990". National Oceanic and Atmospheric Administration. Retrieved 24 December 2012.

"BMC Profile". BMC. Retrieved 20 July 2014.

"BMC takes up wall painting". The Hindu. 12 March 2008. Retrieved 21, February 2013.

"Cities having population 1 lakh and above" (PDF). Census of India, Government of India. Retrieved 2 November 2011.

"City Development Plan: Bhubaneswar" (PDF). JNNURM, Government of India. Retrieved 14 December 2012.

"City Structure" (PDF). Environmental Management Plan for Bhubaneswar. Forest and Environment Department, Government of Odisha. Retrieved 26 December2012.

www.census2011.co.in/census/city/270-bhubaneswar.html. date of visit 20.04.2020

11

River Pollution in India: A Socio-Legal and Sustainable Approach

Dr. Abhijit Bhattacharjee[1] & Soumya Pratik Dutta[2]

[1]Assistant Professor, Jalpaiguri Law College, Jalpaiguri, West Bengal, India

[2]Ph.D. Research Scholar, Department of Law, Assam University, Silchar, Assam, India

Abstract

River is the lifeline of ecology as well as human civilization. River water pollution is a serious threat to human existence. There are various social, economic, cultural and financial reason of river pollution in India. Though the adverse effects of river water pollution are not immediate but it is obvious, absolute and grave. Being a renewable natural resource in general and due to its easy availability, in most of the cases the river pollution issues are compromised. The authors here tried to focus on some rational problems which the rivers in India are facing in regular basis. Only detecting the pollution sources is not enough but the mass awareness, "water pricing" approach and most importantly sustainable use of this river water is the need of the hour. Apart from judicial approach, a specific legislation on protection of rivers of India can also be an effective step towards this end.

Keywords: River, Pollution, Legislation, Sustainable.

[1]Corresponding Author

Introduction

Rivers and streams are often considered major features within a landscape as a major source of drinkable water; however, they actually only cover around 0.1% of the land on Earth. The importance of river in human life cannot be ignored. If water is considered as life then river is one of the major natural sources of the life. Many human cities and civilizations are built around the freshwater supplied by rivers and streams.

Most of the major cities of the world are situated on the banks of rivers, as they are, or were, used as a source of water, for obtaining food, for transport, as borders, as a defensive measure, as a source of hydropower to drive machinery, for bathing, and as a means of disposing of waste as well. Rivers have played a major role in all the great human civilization of the world, whether it was Egyptian Civilization around the river Nile or Mesopotamian Civilization near the Tigris River or the Indus valley civilization. Rivers in India are considered as sacred and they are even worshiped as God or Goddess. In India especially among the Hindus almost every religious ritual is incomplete without involvement of river. Not only that, India has rich resource of rivers and in almost every part of India there is a nearby river. Undoubtedly, rivers in India play an important role in India's economy as India is an agriculture-dominated country and nearly 70% population of India depends directly or indirectly on rivers for their livelihood. It has a major role in irrigation, transportation, fishing and so on. So, its role is not only confined to agriculture or other daily life works but they are tightly bounded to Indian tradition and culture. River is one of the prime reasons for growing up natural forests and wildlife across the world. Amazon rain forest at the bank of Nile river is one of the greatest examples of it. Not only human civilization but the existence of entire flora and fauna on the earth is dependent on the water or in other words rivers. Most of the forests and wildlife as well thus also flourished on the bank of the rivers. So, the importance of the river in real sense cannot be ignored. Moreover, it is the duty of every human being to protect the natural resources of the water in the world, but unfortunately, it's not the wildlife but we, the human beings who are claiming to be the most intelligent creature on the earth, polluting and destroying such vital source of life, i.e., river in the name of development and progress. Being the oldest water body of the earth river has its own biodiversity that is the most diverse and complex in comparison to other ecological systems of the Earth. Rivers play important role for survival of a large variety of flora and fauna. Fish, amphibians, birds, insects, invertebrates, and reptiles live in rivers, or find their food there. Rivers use as the prime medium to link with the food chain for the natural habitats around the river and as such, river biodiversity helps them to extend beyond the surface area they cover. This habitat connectivity role functions by connecting both sides of riverbanks. However, there is a judicially evolved fundamental right to water, such a right is not explicitly incorporated under the Constitution of India. However under the Fundamental duties enshrined in Article 51-A(g), it is stated, "It shall be the duty of every citizen of India to protect and improve the natural environment including forests, lakes, rivers and wild life, and to have compassion for living creatures" so the protection for the sources of water is obviously mentioned in the Constitution of India. The closest that we came to directly incorporating this right was when the National Commission that reviewed the Constitution recommended in its report in 2002 that a new Article 30D be inserted in the Constitution thus: *'Every person shall have the right – (a) to safe drinking water ...'* That recommendation of the National Commission reiterated what the higher courts have been holding in similar words in the past few years.

Definition of River

According to **Wikipedia**, "A river is a natural flowing watercourse, usually freshwater, flowing towards an ocean, sea, lake or another river. In some cases, a river flows into the ground and becomes dry at the end of its course without reaching another body of water. Small rivers can be referred to using names such as stream, creek, brook, rivulet, and rill. There are no official definitions for the generic term river as applied to geographic features, although in some countries or communities a stream is defined by its size. Many names for small rivers are specific to geographic location; examples are "run" in some parts of the United States, "burn" in Scotland and northeast England, and "beck" in northern England. Sometimes a river is defined as being larger than a creek, but not always: the language is vague."

According to **Oxford dictionary**, "river is a large natural stream of water flowing in a channel to the sea, a lake, or another river."

According to **UNESCO**, A river is a course of water that originates in the mountains and flows downwards until it reaches the sea. On its perpetual journey, river water crosses land, hills and plains.

So, we can say A **river is a** large, flowing waterway, primarily carry freshwater which carrying water down from higher areas to a lower point, ending at an ocean or in an inland sea.

Primary Causes of River Pollution

As it is observed from its definition that rivers generally carry freshwater, but due to some riparian as well as anthropogenic activities, the river water is being polluted. There are so many reasons of pollution of river water. Some of the major reasons are discussed as follows:

Oil and Natural Gas Exploration

Understandably, India needs to attain self-sufficiency in energy. Hence, the Ministry of Petroleum and Natural Gas routinely auctions oil and natural gas blocks to Indian companies for exploration in various river basins. While exploration firms claim to take all the necessary steps to prevent pollution, it is but natural that their methods are not 100% efficient. Hence, it is often seen that leakage of oil and natural gases polluted the river slowly and silently. Bhopal Gas Leak disaster in December 1984 is one of the greatest examples of it.

Chemicals & Effluents

Industrialization along the river belt is polluting the water with chemicals and other industrial effluents. While large corporations adopt advanced techniques to treat effluents and chemicals to negate or reduce their toxicity, smaller firms often have no such considerations and violate environment safety standards. Similarly, various hazardous bio-medical wastes that are discharged from various hospitals also polluted the river water either directly or indirectly. It is also pertinent to mention here that at the time of idol emersion in the river the harmful chemicals in the colours of the idol and other accessories also increase the quantity of harmful chemicals in the river water, which not only affects the riverine animals and plants but affects the human indirectly as well. For example, a scientific research revealed that, harmful 'Lead' (PB) found in the 'Boroli" fish of Teesta river in North Bengal.

Garbage Dumping

High population density around the riverbanks and the reckless dumping of non-biodegradable waste is further adding to water pollution. Despite warning and strict fines imposed by local administrations on those found dumping garbage into rivers and their estuaries, this uncivilized practice continues unabated. It is often seen that local fish and meat markets are mostly situated at the banks of the river, so, the waste animal body parts and other garbage are easily thrown away in the river. Likewise, plastic wrappers, household waste etc. also recklessly throw into the nearby river water. Due to lack of proper legal authority, these silly human activities are now growing day by day.

Washing and Sewage

Laundering clothes on river banks is a common sight in India. Modern detergents are made of chemicals that contaminate river waters. Defecating around river banks is yet another horrible practice in rural parts of the country that contributes to rising pollution of rivers in India. The situation is worsened by local administrations and private operators of sewage collection tankers dumping the waste in rivers and their estuaries. Apart from it in some places of India river banks are used as cow or bull shed and obviously the river water there become an ideal place for bathing of those animals and as a result the river water getting polluted.

Cremation and Last Rites

Devout Hindus attach a lot of significance to life after death. Hence, cremation grounds in rural India are located on the banks of rivers. Unfortunately, many Indians families that cannot afford cremation consign mortal remains of loved ones to rivers. These human cadavers cause severe water pollution. Especially in the cities all the garbage arose from such cremation rituals are indiscriminately dumped into the river. A strict step from the local self-government, i.e., Municipalities or Panchayets is the need of the hour. It is their duty to regulate the cremation procedure sustainably without compromising the health of the river.

Sand Dredging

High-quality sand from riverbeds is needed for India's booming construction industry. This has led to rising pollution in Indian rivers. Dredging operators- usually the unlicensed ones- deploy kerosene and diesel-fueled cheap watercraft manned by cheap labour for sand dredging. Engines of these boats cause pollution. Sand dredging or quarrying has now become a profitable business for some businessperson. In the name of business, they often violate the permissible quantity of Sand quarrying as prescribed by National Green Tribunal across India. Especially in North Bengal region such illegal sand quarrying is now become a major cause of concern for the police administration well. Due to such activity's river causes soil erosion, even sometimes during heavy rainfall it changed its natural direction and destroyed many human lives and properties.

Plastics and Thermocol

Apart from the above causes indiscriminate dumping of plastics and Thermocols, which are not bio disposable and very harmful for the entire river biodiversity are one of the serious threats to not only for the rivers but the human civilization as well. These are emerged as harmful pollutants in a very recent past. However, the continuous and rapid use of it definitely posed an irreparable damage to the entire ecology. Like marine biodiversity in the rivers as well a number of riverine animals including turtle and fish died due to consumption of such plastics and thermocols. It also obstructs the natural flow of the river. They also played a major role in contamination of river water as well.

Agriculture Waste

In many places of India River is the only source of fresh water, which is suitable for agriculture and the corps, grow into the adjacent land of river bank due to fertile soil. However, it is seen that the pesticides, insecticides and other harmful wastes, used for the agricultural purpose often mixed with the river water through interlinking channels between the agriculture field and the river. This way the river water is polluted through agricultural activities.

These are some of the main reasons for rising pollution of rivers in India. Unfortunately, this contamination of Indian rivers- revered as holy by the indigenous community- has a severe environmental impact that affects humans too.

Possible Consequences of River Water Pollution

Rising pollution of rivers in India is affecting humans as well as other flora and fauna. It has a direct impact on the various features of this country.

Impact on Flora and Fauna

Chemical, effluents and sewage that pollute Indian rivers is causing several species of aquatic life to go extinct or move away to safer havens. This has a direct impact on wildlife conservation efforts exerted by the Central and state governments at national parks and wildlife sanctuaries. River contamination threatens biospheres and nature conservation areas. Migratory birds shun these rivers, and hence, they can face extinction. Other species of flora and fauna are threatened by loss of habitat caused by rising pollution of rivers in India.

Loss of Livelihood

Fishermen and fish farms that once flourished on banks of various Indian rivers are finding it increasingly difficult to find sufficient catch of edible fish. Such aquatic life forms have moved away from polluted waters or may face extinction. Fish from polluted rivers is found to be high in mercury, lead and cadmium and hence, unfit for human consumption. Also, edible fish is contaminated with Salmonella, Shigella and other harmful microbes found in human faces. Thus, such fish is unfit for human consumption.

Drinking Water

According to Niti Aayog, which replaced Planning Commission, whopping 200,000 people lose their lives every year due to various problems caused by consuming contaminated water. The health hazards associated with infected water range from cancer to gastrointestinal disorders that occur due to deadly microbes that are creeping into the rivers. Over 800 million Indians are severely stressed for potable water due to river water pollution, among other reasons. This can severely stunt India's economic growth.

Agriculture

While agriculture is impacted adversely by rising pollution of rivers, it is also one of the reasons for contaminating waters. The use of modern pesticides, herbicides, and fertilizers is polluting our rivers. Interestingly, this impacts the overall productivity of the agricultural land as river water contaminated by industrial effluents cannot be used for irrigating crops. Polluted water does not allow seeds to germinate and cause stunted growth, denying farmers of a bumper harvest. Indigenous agriculture is vital to India for maintaining self-sufficiency in food.

Disease and Health Burden

Rising river water pollution in India is a significant cause of concern since it can cause moderate to severe health conditions among humans and domestic animals. Bathing in contaminated river waters causes skin diseases, allergies, and other such ailments. Consuming polluted water can cause cancer, depletion of calcium from bones of humans and animals (Osteoporosis), loss of vision, impotence among men, sterility among women, Tuberculosis and other severe medical conditions. Majority of Indians depend upon state-sponsored healthcare system since it is readily available and affordable. Hence, state governments, union territory administrations, as well as the central government, face a major health burden that translates into billions of Rupees spent towards providing medical services to those afflicted.

Loss of Export Revenue

Freshwater fish varieties including the famous Hilsa, Rohu, Katla and prawns from Indian rivers once had a high demand in foreign countries, especially in the Middle East. Sadly, river water pollution has caused these varieties of prized fish to get contaminated with disease-causing microbes and chemicals. Consequently, several countries have banned imports of freshwater fish from India, including farmed varieties. This causes a severe loss of export revenue for India.

River Pollution and Sustainable Development

Our current life is totally dependent on rivers. The river systems provide irrigation, potable water, cheap transportation, electricity, as well as livelihoods for a large number of people all over the country. Water, once considered to be a renewable resource, is gradually becoming a non-renewable resource. Hence, current water management practices warrant a new and conservative approach. Water, a natural capital, generates a variety of heterogeneous ecosystem services. Today, India's Rivers, riverine biodiversity and river dependent communities are facing major threats: from large dams, pollution, encroachment, sand mining, deforestation and bad management practices. These factors are affecting all aspects of rivers: ecological, social, cultural, religious, aesthetic, tourism-related and economic. Rivers face many existential threats from dams (including run-of-river and other hydropower projects), pollution, diversions, encroachments, excessive use of groundwater and catchment degradation. There are a number of laws and related institutions (for example, River Boards Act 1956, Water Pollution Control Act, 1974 and the State and Central Pollution Control Boards existing since 1974) that have remained ineffective and problematic. The Government unfortunately looks at rivers as a resource to be exploited for various water services, and not as a resource, that has certain basic economic, social, environmental and cultural value. It does not mean rivers cannot be used for developmental purpose; but the concern is development should not be at the cost of the life or destroying natural flow of the river. The conservation and development of the river should go hand to hand. Rivers water usage should be planned in such a way that it meets the basic needs as defined in the Constitution of India, i.e., right to life. For that right to be safeguarded this means that maintaining the health of the rivers needs to be maintained i.e., ensuring minimum environmental flow in all the rivers, and also meeting the growing water demand for economic development. Natural flow of the rivers should be given the first priority. Unfortunately, there is no specific law in India that requires that when such projects are taken up the projects must ensure that rivers continue to flow with perennial freshwater flow. Moreover, there is no legal or institutional mechanism to ensure that the assessment process, the planning process, the decision-making process, the cost benefit analysis or the environment impact assessments include the cost of the destruction of the existing resource of rivers and destruction of the services provided by that resource. What this thus implies is that the official system has absolutely no value for the river as a resource. The only state in India that has a clear policy on this is Himachal Pradesh (HP). In the absence of necessary legal and institutional mechanisms to ensure implementation of river friendly aspects of the existing policy it is not possible for the existing Boards and Authorities to tackle all the issues relating to rivers in India as because the nature and characteristics of the rivers in India is different from one another. Only awareness campaign among the riparian and general people is not enough so a separate central enactment is urgently required for the protection of rivers in India.

Conclusion

In this context, it is worth here to mention that, in June 2014, the national mission towards a clean Ganga was launched. With a budget of Rs. 20,000 crore, **'Namami Gange'** Programme aimed to accomplish the twin objectives of effective abatement of pollution and the rejuvenation of the Ganga. Its implementation proposed a 10-year plan and set forth a number of activities, to target both short-term and longer-term impacts. In April 2017, a draft was submitted to Minister for Water Resources, River Development, and Ganga Rejuvenation Uma Bharti, for the proposal of a new piece of legislation, the *National River Ganga (Rejuvenation Protection and Maintenance) Bill Act 2017*, mainly to ensure the cleanliness and rejuvenation of the river. The proposed Ganga Act seems to be in its final stage and, if successfully passed, will make it the first river in the country to be protected by an Act of Parliament. If successful, the Act could help protect the river in a number of ways. Not only it would help to establish the rivers rights to exist and evolve it would also establish the rights of people and the environment to have access to a clean and health water source. Since enforcement of provisions of the Water Act, 1972 and Environment Protection Act 1986 is under the purview of local civic bodies in respective states, the Environment Ministry plans to set up sewage management systems with private participation. Meanwhile, the Central Pollution Control Board has identified 302 stretches along 275 rivers in the country that are polluted, based on the Biochemical Oxygen Demand (BOD) - a critical parameter of water quality. It is the amount of dissolved oxygen in water, used in meeting the metabolic needs of aerobic microorganisms in water having organic pollutants and sewage. This is primarily due to the discharge of untreated municipal wastes and industrial effluents into the rivers. Maharashtra has the highest number of polluted rivers stretches at 49, followed by Assam with 28. In Delhi, there is one river, and that is polluted - the Yamuna. Agricultural prosperity of Bengal has been intricately linked with its rivers. Govt. of India has fixed the bathing standard in flowing water. Where dissolved oxygen is 5mg/litre, maximum Biochemical Oxygen Demand is 3mg/litre and bacterial count is less than 500/100ml, the water is treated as fit for bathing. Unfortunately, no river in West Bengal meets the noted standard. The river and groundwater should be treated as common property resource (CPR). The management of rivers and groundwater will be impossible without peoples' participation as there are strong traditions of community management of water and other natural resources in India. The traditional systems got marginalized as the State became powerful and came out with a new governance and management system of centralized command and control. The need of hour is to learn from the past and develop a new model that can safeguard the health of all the natural resources and ensure equity in access to them by all section of the society. Last but not the least, Rivers are the property of society, i.e., the people who live in the country, and not of any individual or business. The National River Policy should act as guide in making the rivers an entity owned by all and the state should act as custodian and not owner.

References

Central Pollution Control Board (CPCB) (2013). "Status of Water Quality in India." A Report by Central Pollution Control Board, Ministry of Environment and Forests, Government of India.

Hudda Sonakshi, *Editor*, Janhit Foundation, Meerut; "River Pollution: Causes, Actions and Revival, For a better tomorrow, act today", Systems Vision A-199, Okhla, Phase-I, New Delhi - 110 020.

Major General (Retd.) Chaturvedi, Ajay Kumar AVSM, VSM; "River Water Pollution - A New Threat to India: A Case Study of River Ganga"; Vivekananda International Foundation 3, San Martin Marg | Chanakyapuri, New Delhi- 110021.

Murlidhar, S. (2006). "The Right to Water: An Overview of the Indian Legal Regime." In E. Riedel and P. Rothen (Editors), "The Human Right to Water ". Berlin: Berliner Wissenschafts- Verlag.

Shah, Natasha; an article on "Explained: Causes and Impacts of River Water Pollution in India", Youth Ki Awaz on 1st November, 2018.

Tripathi, B. D., and Tripathi, S. (2014). Issues and Challenges of River Ganga. In Our National River Ganga. Springer International Publishing.

UNESCO. (2015). *"Water for a sustainable world"* (The United Nations World Water Development Report 2015). Paris: Author.

12

Covid-19 Pandemic in India: A Geo-Environmental Analysis

Biswajit Majumdar

Assistant Teacher, Fatepur C.D. High School, Kaliyaganj,
Uttar Dinajpur. West Bengal, India

Abstract

Covid-19 is a viral infection disease; which has created a horrible picture of present situation of mankind. It has created the dilemma between mankind and environment. During COVID 19 period Indian Government has decided Lock Down whole country. This lock down period environments are healing gradually. Air Quality Index is very sweet able in Indian metropolitan cities. After 30 years of Ganga Action Plan water quality of Ganga river have improved 40% during lock down time. COVID 19 has shown a skeleton of Indian labour migration. Many labours have lost their lives during back way to their home. India has a cultural and ethnic diversity, so it helps to protect from spreading this viral infection. Indians have some genetic advantage to boost the immune system. During lock down period Indian Government has ensured the food stuff of every Indian with free of cost, though in this period Indian GDP is falling down gradually. This decision has shown that India played a Well Fare state. But COVID 19 has emerged some question of geographical studies. Which will be creates a new Paradigm of studies.

Keywords: Lock Down, Healing Environment, Industrial Effluent, Internal Migration, Government Policies.

Introduction

Coronavirus is a large family of viruses which may cause illness in animals or humans. In humans several corona viruses are known to cause respiratory infection from the common cold to more severe diseases such as middle east respiratory syndrome and sever acute respiratory syndrome. The most recently discovered corona virus causes corona diseases COVID 19. This new virus and diseases were unknown before the out brake in Wuhan, in China, in December 2019. It is a Zoonotic virus, which means these viruses are able to be transmitted from animal to humans. On the other hand, it called Novel Corona virus because it is a new corona virus that has not been previously identified in humans. Covid-19 is the most recent to make the jump to human infection. It spreading from one person to others via droplets produced from the respiratory system of infected people, often during coughing or sneezing. The full form of COVID 19 is Corona Virus Disease of 2019, this abbreviate name was taken by WHO, when they are press release on February 11, 2020.To protect from the pandemic of COVID 19 like whole world Indian Government has decided Lock Down in Whole country from 24th March 2020.

Covid-19 Pandemic in India

The first case of Covid-19 il1ness in India was reported on 30th January 2020. On 22nd March India observed 14 hours voluntary public curfew at the instance of the Prime Minister of our country. It was followed by mandatory lockdowns in Covid 19 hotspots and all major cities. Further on 24th March the P.M ordered a nationwide lockdown for 21 days, it further extended to 31st May. From 1st June,2020 the Government started "unlocking" the country in various unlock phases.

Impact of Covid-19 in India

Indian Government announced complete lockdown for 68 days in Four phases, Which period creates some impact of socio economic and environmental aspects which are as follows:

Changing Environmental Conditions during Lock Down

India is a second most populous country in the world. Total population multiplied by more than 3.5 times with in last one hundred years i.e.

1901 to 1991 and 4 times from 1991 to 2020, Which simply means that the pressure on Indian's natural resources has increased about four-fold. During pandemic, Indian Government has decided Lock Down whole country. During this lock down, after 10 years blue sky appeared and some environmental phenomena has shown in the sky of Delhi. Though we are known that Delhi is one the most polluted cities in the world. The healing environmental phenomena is:

(a) New Delhi Govt data shows the average concentration of P.M 2.5 Plunged by 7.5 % in the space of a week falling from 91 microgram per cubic meter on march, 20 to 27, after the lock down began. (b) Nitrogen dioxide went from 52 per cubic meter to 15 PCM in the same period also a 71% fall. (c) During the first three weeks of march, the average nitrogen dioxide levels declined by 40 to 50% in the cities of Mumbai, Pune and Ahmedabad. (d) Dangerous pollutants P.M 2.5 and the large P.M 10 which are less than 10 micrometers in diameter also dropped steeply.

Table 1: Delhi Air Quality Index

2020		
	18th Feb	18th June
A Q I	308	129
P.M- 10	260	59
P.M - 25	130	144

Source: w.w.w.timesnownews.com.ind. 18 June 2020. 8.45 AM. IST.

Healing The Ganga River

Before and after lockdown 36 real time water quality monitoring stations in the river Ganga. Sea wage totaling about 3500 million liters per day is discharge into Ganga, of which only 1100 MLD is treated discharged. About 300 MLD of it is industrial effluent comprising about 9 % of total waste water discharged into river daily. The question is which are significant sources of pollution in Ganga? The answer is the Industrial Effluents. After lockdown Ganga water in Haridwar becomes "Fit to drink" after decades. According to Uttarakhand, Environment Protection and Pollution Board, there has been a 34% reduction in fecal coliform and 20% in Biological Oxygen Demand in Haridwar. Such a remarkable improvement has not been witnessed in the past 30 to 40 years. According to Uttar Pradesh Pollution Control Board, healing water should have a Dissolved Oxygen Level of at list 7 mg/liter. The Dissolved Oxygen Level up stream in the river Ganga is 8.9 mg/ liter while in the downstream it is 8.3 mg/liter. This clearly shows that water quality has improved significantly and is optimal for bathing. It has been seen that about 40 to 50% improvement in the quality of water in Ganga. Due to lock down, 22 drains which disposed of sea wage into the river, have been sealed too, hence making the water cleaner. According to an estimate about 19,65,900 kilograms of polluted matters are discharged into the Ganga every year. It has been seen that one of the most polluted part has gone through U.P and West Bengal.

Table 2: Status of Ganga River Pollution

Status of Ganga River Pollution		
State	**2005**	**2020**
Uttarakhand	3mg/< BOD	< 3mg/< BOD
Uttar Pradesh	7-8mg/<BOD	1.25-402 mg/<BOD
Bihar	2.4 mg/<BOD	< 3mg/<BOD
West Bengal		2.3-401 mg/<BOD

Source: National Mission for Clean Ganga.

Bio-Diversity Impact During Lock Down

The comprehensive study of bio diversity is very important for the conservation and increase of bio diversity of a nation because the status of bio diversity determinates the health and wealth of nations. Bio diversity threat is a present problem of environment but during lock down this situation is reversed. After 15 to 20 years Dolphins are coming back around Mumbai coast. Additionally, many videos and photos of animal presence during lock down have in fact turned out to be cases of misinformation and unrelated to the lock down the elephant, deer, whales, peacocks etc. During lock down increased visibility which stimulate to came out them from jungle.

On the other hand, a traffic analysis finds a significant increase in reported poaching of wild animals in India during the lockdown period that is not restricted to any geographical region. Reports of poaching incidence for consumption and local trade have more than doubled during lockdown although there was no evidence of stockpiling of wildlife products for future trade.

Immunity System of Indians

India has cultural and ethnic diversity; it helps to protect from spreading this viral infection. Otherwise Indians have some genetic advantage because they have lived non hygienic environment. So, they have evolved to gain more genes that protect against viral infections. Two families of genes KIR genes and HLA genes play a part in this protective functions. Indians have more KIR genes than the Chinese and Caucasians. This could make Indians more immune to the virus.

On the other hand, BCG vaccine, various nutritional programs such as Applied Nutritional Program, Special nutrition, Balwadi nutrition, Supplementary feeding programs, Vitamin A prophylaxis program boost the immunity system of Indian at large.

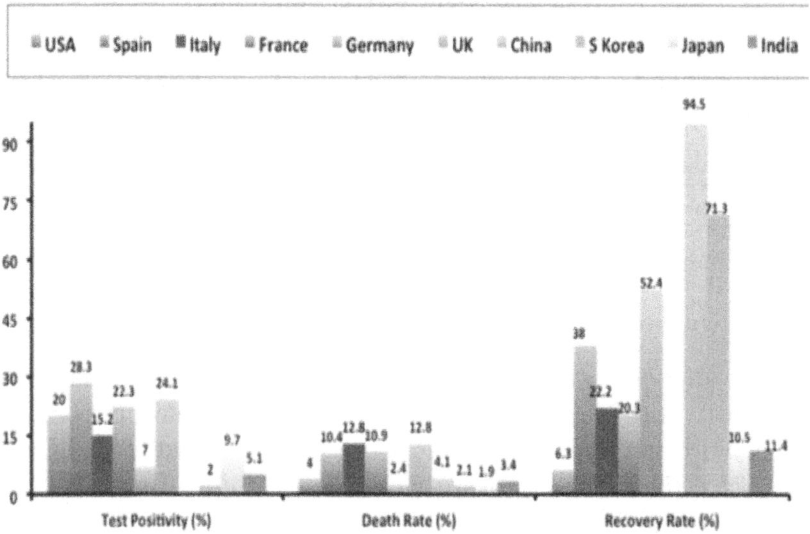

Figure 1: Comparative Analysis
Source: www.immunopaedia.org.za.17th April. 2020. 8.45 AM IST.

Situation of Migrant Labourer During Lockdown

India is a developing country, so laborers migrate from rural areas to urban areas. Even it is a much more disparity between the states labour force. Covid-19 has shown the horrible picture labour migration. The skeleton of Indian labour class life has open to all over the world. According to World Bank Report lock down in India has impacted 40 million internal migration. Around 50,000 to 60,000 moved from urban centre to rural areas of origin in the span of a few days. Many labours have lost their lives during back to their home. They are back to their home in path way but no trains and bus service provides for their facilities. There is much more politics has been shown for the migrant labour. Contrast these tragedies with the recent dilution of labour laws in three states (U.P, M.P, and Maharashtra) to push forth economy recovery. The cumulative picture is of a blatant disjunction between the reality of the worker migration and the policy adopted to tackle the crisis of "the labour question." Ministry of labour and Employment Government of India have decided some advisories on Covid 19 as (a) Extension of return filling (b) Region wise detailed of Covid 19 control rooms. (c) Ministry of Labour and Employment advises all public and private employers to support their employ and workers. (d) Preventive measures to contain the spread of Covid 19. (e) Established 20 Grievance centre.

Table 3: Migrant Workers During Covid-19

Migrant Workers During Covid-19	
States	Worker
Kerala	302710
Maharastra	50000
Gujrat	42000

Source: www.economictimes.indiatimes.com. 23April. 2020. 9.27 AM IST.

Indian Government Policies During Covid 19 Pandemic

India announced a Rs 1.7 lakh crore relief package to take care of poor workers and those who need immediate help amid a lockdown to combat the coronavirus pandemic. The Pradhan Mantri Garib Kalyan Scheme, which will cover 80 crore poor people, will include cash transfers and food security. Each one will get free 5 kg of rice or wheat every month for the next three month. The scheme also focuses on giving payments to different sections of the society directly into their bank accounts. This will cover farmers, workers under the Mahatma Gandhi National Rural Employment Guarantee scheme, widow pensioners, disabled, women with Jan Dhan accounts, women running self-help groups, beneficiaries of the Employees Provident Fund Organization and construction workers.

The 10 points of relief packages are:

1) Payout given 8.69 farmers under PM KISAN will be front loaded.
2) Wages under MGNREGA to be hiked to Rs 202 from Rs 182.
3) Three crore poor senior citizens, widows, disabled to get one-time ex gratia amount of 1,00 in two instalments.
4) 20 crore women Jan Dhan account holders will get an ex gratia amount of Rs 500 per month for three months.
5) Women in 8.3 crore BPL families covered under Ujjwala scheme will get free cylinders for three months.
6) Doubling collateral free loan amount to Rs 20 lakh for women in self-help groups.
7) GOVT will pay PF contributions for both employer and employee for three months.
8) PF fund scheme regulations to be amended to allow non-refundable advance of 75 % of amount in account or three months of wages, whichever is lower.
9) To direct the State Government to utilize Rs 31,000 crore under the Building and other construction Workers Fund to support 3.5 crore construction sector workers.
10) The Government will provide medical insurance cover of Rs 50 lakh per person to those at front lines of fighting the virus. This would cover doctors, nurses, paramedical staff, sanitation and ancillary workers.

Despite this scheme Indian Government has launched the Arogya Setu app to facilitate Covid 19 Patients. Otherwise Indian Government has established a special ward/hospital for Covid 19 patients.

Conclusion

Geography is what geographers do, because the term 'geography' means, and has meant, different things to different people in different times and places. Geography is a dynamic discipline, not static. So, from the geographical point of view Covid 19 has opened our eyes to analyze some impact of geo-environmental studies. Nitrogen dioxide drastically change during lock down period. Surface Ozone has increased in whole world. On the other hand, Ozone Hole are gradually smaller than the other year. Ganga Action Plan had been taken in 1986, but during lock down in three months Ganga river much cleaner than the 30 to 40 years. Dolphin mammals have much more increased in Mumbai coast, it indicates that environment has gradually healing during lock down effect. Air Quality Indexes are higher in Indian metropolitan cities during lock down time. A non-hygiene environment can protect against viral infection, but it is not a good sign for mankind. BCG vaccines are banned in developed countries. But in India it has become game changer. Indian Government Policies during lock down period has shown that India Played a role of a Welfare Nation. But according to the Small Farmer Development Agency Indian Government has taken 100 days' work to stop the migration of labour during this time. Covid 19 has proved that there has been disparity of wages and work force among the Indian states. Some experts analyzed the lockdown decision in India. On the other hand, although Covid 19 was first traced in December but WHO officially declared it as a pandemic on march 11,2020. Countries experiencing the greatest number of cases include China, Iran, Italy and the Republic of Korea.

References

Choudhary, Sristi. (29 April 2020). Lockdown: Ganga flowers cleaner though just marginally.w.w.w.livemint.com.ind.8.30.AM IST.

Johanston, R. J. and others., The Dictionary of Human Geography, USA: Blackwell Publishing., 2005.pg 304.

Labour welfare, Ministry of Labour and Employment, Government of India, www. labour.gov.in. 5 th June.10.45 A.M. IST.

P.T.I. (23 April 2020). Lockdown in India has impacted 40 million internal migrations: World Bank.w.w.w.Economictimes.com.9.27 AM IST.

Shukla, Nelanshu. (22 April 2020). Lockdown effect: Ganga water fit for drink after decades. w.w.w.The new India express.com.2.53 AM IST.

Singh, Savindra, 2008: Environmental Geography: Prayag Pustak Bhawan. Allahabad. Pg. 504-505.

Sinha, Nitin. (15 May 2020). Perception, Legality and Politics of the migrant workers crisis in lockdown. www.The wire. in. 9.15 P.M.IST.

Tandan, Aditi. (3rd March 2020). India lockdown: Reports of reappearing animals are not just misleading they are also harmful. www. scroll. in.1.30 P.M. IST.

Upadhay, Vineet. (8 May 2020). Lockdown effect: Ganga water fit for drinking after decades, www.the new india express.com; 2.53 a.m IST

Varshney, Vibha. (9th April 2020) Covid-19 Do Indians have higher immunity r novel corona virus. www.downtoearth.org.in 8.00 p/m IST.

Wright, Rebecca. (1st April 2020). Indias corona virus lockdown is having11 a dramatic impact.w.w.w.cnn.news.com. 6.56 AM. IST.

13

Changing Paradigm in the Educational Environment During COVID-19 Pandemic: An Indian Perspective

Dr. Dipak Kumar Singh[1] & Dr. Ranjan Sarkar[2]

[1]Assistant Professor, Department of Education,
Ananda Chandra Training College, Jalpaiguri, West Bengal, India
[2]Assistant Professor, Department of Geography,
Ananda Chandra Training College, Jalpaiguri, West Bengal. India

Abstract

The present paper is an attempt to analyse the changing paradigm in the field of education during COVID-19 pandemic in Indian context. The humanity is undergoing through a critical phase of uncertainty, unreliability and unpredictability. COVID-19 pandemic has brought tremendous changes in every aspect of life including education, entertainment and communication. The world of education is facing severe and several challenges throughout the world. In Indian context challenges are different from the rest of the world due to the unique, diverse and complex nature of Indian subcontinent. Education around the globe including India is undergoing through a paradigm shift. Technology has played a critical, decisive and significant role in this changing paradigm. Traditional forms of face-to-face learning have been entirely replaced by virtual learning, digital learning and online learning during the lockdown of educational institutions to

[1]Corresponding Author

combat COVID-19 pandemic. This situation has created numerous opportunities as well as challenges to the learners as well as teachers. In India digital divide is one of the significant challenges in the modern methods of teaching learning process.

Keywords: COVID-19, Lockdown, Online Learning, Digital Learning, E-Learning, Digital Divide, Virtual Reality.

Introduction

The world is going through a critical, uncertain and unpredictable stage of disaster, devastation and destruction because of COVID-19 pandemic. India is not an exception. All the aspects of life are getting severely influenced including education, entertainment and communication.

Education system throughput the world is going through a paradigm shift. The role of technology is crucial, critical and decisive in this paradigm shift. Radical changes are being seen in the field of education in the form of digital technology. Indian education system is also undergoing through tremendous changes. All of a sudden new form of learning like virtual learning, online learning, digital learning etc. replaced the traditional forms of face-to-face learning because of unavoidable circumstances during COVID-19 pandemic. To mitigate the spread of this pandemic policymakers in the field of education are encouraging online initiatives as much as possible. These online initiatives have changed the face of Indian traditional education system to a great extent. This situation has created abundant opportunities in the world of education. But at the same time a number of severe challenges as well are being faced by the students, teachers and parents. Since the options are limited during COVID-19 pandemic, teachers and learners are bound to go through online activities to continue the educational process as much as possible. The virtual platforms of learning have given a hope to the teachers and learners during this pandemic.

India is country of diversities, differences and divergences. In Indian context nature of challenges being faced during COVID-19 pandemic are different from the rest of the world to a certain extent. A large section of Indian population is still below the poverty line. The children belonging to marginalised and underprivileged section of the society are dependent on schools not only for education, but also for food. Closure of schools to mitigate the impact of COVID-19 pandemic has created so many difficulties to these children. These children are unable to access the online education due to lack of awareness, parental support, internet, infrastructure and basic amenities. They are worse affected. Policymakers are worried regarding these issues. But they have very limited options to carry out the education of the children during COVID-19 pandemic situation. Online education, digital learning, virtual classrooms etc. are the few options by which education of the children can be continued. Internet connectivity plays a significant role in the process of online teaching learning process. Children living in remote regions face so many problems concerned with internet connectivity and consequently they suffer a lot. This situation stimulates to think seriously regarding the issues faced by these children. The goal of universalisation of education cannot be achieved without resolving these issues.

Indian Education System During COVID-19 Pandemic

The global spread of COVID-19 pandemic has changed everything. All the spheres of human life got tremendously affected. Education is a significant activity which is necessary for the growth and development of human race. Education at all levels in India has been severely affected by the spread of the COVID-19 pandemic. School learning, face-to-face learning and other traditional forms of learning were completely stopped all of a sudden to combat this pandemic. This situation has created a deep sense of uncertainty and unpredictability in the hearts and minds of teachers, learners and parents. Most of the educational institutions in India are closed to mitigate the spread of COVID-19 pandemic. How to overcome the impact of this pandemic on the education system is one of the greatest challenges in front of the educationists, administrators and policymakers. Technology has played a significant and pivotal role in Indian education system during this pandemic.

The use of digital technology is being seen as an alternative to the traditional face-to-face learning during the COVID-19 pandemic. But in Indian context digital divide is also a harsh reality which can act as a great hindrance in the implementation of digital learning specifically in remote regions (Antonio & Tuffley, 2014). A large section of Indian society is still below poverty line. This section still does not have the access to the basic amenities of human life. The difficulties and problems in the life of the people belonging to this section got further multiplied during COVID-19 pandemic. Moreover, children of families below poverty line were dependent on schools not only for education, but also for food through Mid-Day Meal Scheme of the government. But closure of schools during COVID-19 pandemic has made the life of these children more pathetic. In this situation the idea of virtual classroom, digital learning, online platforms etc. seems irrelevant in case of these children who are not able to fulfill their basic needs.

Transformation in Indian Education System

Transformation is a continuous and dynamic process. It plays a significant role in the process of growth and development. It is an integral part of any evolutionary or revolutionary process. The nature, design and pattern of education always get transformed according to the needs, desires and aspirations of individuals, societies and nations. Technological advancement and digital revolution are compelling the education system to get transformed (Lou, Abrami, & d'Apollonia, 2011). Further, the spread of COVID-19 pandemic and closure of educational institutions intensified this process of transformation. Indian education system is going through a tremendous transformational change where traditional forms of learning have been entirely replaced by modern methods of learning to combat COVID-19 pandemic. The process of teaching, learning and evaluation is getting dependent on technology and its uses. On the other hand, over use of technology and the issue of digital divide have created so many challenges as well to the teachers, learners and parents. That is why many educationists are in favour of blended learning which includes online as well as offline mode of learning.

Expanding Role of Technology in Education

Technology has penetrated each and every sphere of life. Modern lifestyle without technology looks like impossible in the present context. The dependence of life on technology is increasing day by day. Education as an integral part of human life is also getting dependent on technology to a great extent (Waxman, Lin, & Michko, 2003). The traditional methods of learning are being replaced by modern techniques and technologies. Technology oriented teaching learning process has brought revolutionary changes in the field of Indian education system specifically during the period of COVID-19 pandemic. During this pandemic period options are very limited. Online initiatives give a hope in fighting against COVID-19. Since technology has entered into every aspect of education, it has evolved as an essential, integral and inseparable part of teaching, learning and evaluation process of modern education system (Bolhuis, 2003). Technology makes the process of teaching and learning more effective, creative and productive. That is why various kinds of educational curriculums at different levels are giving central position to the technology. Technological advancement has entirely changed the face of education. Different platforms of digital media and video conferencing have emerged as powerful tools of teaching learning process in the present era of information and communication technology (Stoller, 2013).

Virtual Classroom as a Reality

The field of education, at all the levels, has undergone tremendous changes during COVID-19 pandemic. Technological advancements have converted the idea of virtual classrooms into a reality. A virtual classroom is generally an online learning environment that allows the interaction between the teacher and learner. The traditional classrooms are gradually being replaced by virtual classrooms. Further, the spread of COVID-19 pandemic has intensified this process of replacement. Customary methods of face-to-face teaching learning process have been substituted by online teaching learning process during this pandemic. Google classroom, Google Meet, Edmodo Learning Management System, Microsoft Team, JioTeam, Zoom, social media sites etc. are the common platforms which are converting the idea of virtual classroom into a reality. Virtual classrooms are playing a significant role in the life of learners specifically during COVID-19 pandemic. They have the potential to make the teaching learning process creative, interactive and collaborative (Higgins, Xiao, & Katsipataki, 2012). They have given a new hope to the teachers, learners as well as parents. They may play a decisive role in the universalisation of education in the coming future as well. Educationists, administrators and policymakers are well aware of the significance of virtual classrooms in the implementation of right to education.

Digital Learning as an Urgent Need

Digital revolution in India has converted the idea of digital learning into a reality for the common man. Digital learning is the need of the hour. Further, the spread of COVID-19 pandemic has made the digital learning as the urgent need of the hour. Digital learning is a technology based approach of learning (Saeed, Yang, & Sinnappan, 2009). It is providing a safe and secured mode of learning to the learners during this pandemic. It gives a hope to the learners, teachers, parents, administrators and policymakers to continue the education at different levels during this pandemic period. Digital media, digital platforms and digital awareness are playing a significant role in the process of digital learning, self-learning and lifelong learning (Higgins, Xiao, & Katsipataki, 2012). Digital learning makes the process of learning more effective, creative and productive through continuous cooperation, communication and collaboration. It gives opportunities to the learners to learn in accordance to his or her own time, pace and place. It makes the process of learning flexible, interactive and collaborative. Different platforms of digital learning give innumerable opportunities to the learners to exchange, express and communicate his or her ideas, information and imagination with others at global level freely and fearlessly.

Conclusion

The humanity is going through a phase of ambiguity, insecurity and unpredictability. The existence of human life is in danger because of global spread of COVID-19 pandemic. Tremendous changes have been observed in each and every field including the field of education because of this pandemic. Traditional methods of teaching, learning and evaluation are gradually getting replaced by modern means, methods and techniques. This transformation has led to a paradigm shift in the field of education. Technology has played a significant role in these drastic changes. It has penetrated every aspect of human life. Because of this technology education is undergoing through incredible changes which has given birth to abundant opportunities along with numerous challenges as well. Digital divide is one of the significant challenges in technology oriented modern methods of teaching learning process. In Indian context, even after several challenges technology has given a new hope to the learners, teachers, parents, educationists and administrators to continue the education during the COVID-19 pandemic, when most of the educational institutions are physically closed.

References

Antonio, A., & Tuffley, D. (2014). The gender digital divide in developing countries. *Future Internet, 6*(4), 105-108.

Bolhuis, S. (2003). Towards process-oriented teaching for self-directed lifelong learning: A multidimensional perspective. *Learning and Instruction, 13*(3), 327-347.

Higgins, S., Xiao, Z., & Katsipataki, M. (2012). *The Impact of Digital Technology on Learning: A Summary for the Education Endowment Fund*. EEF, London.

Lou, Y., Abrami, P. C., & d'Apollonia, S. (2011). Small group and individual learning with technology: A meta-analysis. *Review of educational Research, 71* (3), 449-521.

Saeed, N., Yang, Y., & Sinnappan, S. (2009). Emerging web technologies in higher education. *Educational Technology & Society,* 12(4), 98-109.

Stoller, E. (2013). Our shared future: Social media, leadership, vulnerability, and digital identity. *Journal of College & Character,* 14(1), 5-10.

Waxman, H. C., Lin, M. F., & Michko, G. (2003). *A meta-analysis of the effectiveness of teaching and learning with technology on student outcomes.* Learning Point Associates, Illinois, US.

14

River Bank Erosion Induced Human Displacement and its Problems: A Study of Maynaguri Block in West Bengal, India

Dr. Madhusudan Karmakar

Assistant Professor, Department of Geography, Maynaguri College, Jalpaiguri, West Bengal, India

Abstract

Bank erosion is a dynamic geomorphic process of a river. Free-flowing rivers tend to reach a state of equilibrium by a process of erosion and deposition. Erosion at one location is roughly balanced by deposition at another location. The bank erosion of the river often causes severe damage to the embankments, grasps fertile lands and displaces a number of people. The main objective of the present study is to identify the spatial distribution of bank erosion and the displacement scenario of people of Maynaguri block in West Bengal, India. The findings of the present study show that Maynaguri block is vulnerable to river bank erosion. River Teesta and Jaldhaka are responsible for this erosion. These two-river cause severe bank erosion for which many people are displaced. Cultivated and habited areas are also lost due to this hazard. To minimize the severity of the impact and the damage potential mitigation measures like structural and non- structural measures have to be adopted in the present study area.

Keywords: Global Human Displacement, Spatial Distribution, Vulnerable Areas, Displacement, Measures.

Introduction

River bank erosion is a recurrent natural geomorphic process which occurs in all channels as adjustments of channel size and shape are made to convey the discharge and sediment supplied from the stream catchment. Bank erosion is the wearing away of the banks of a stream. When river channels are altered under naturally dynamic hydrologic conditions, the river readjusts itself with respect to dimension, profile and pattern to reach its former balance or equilibrium (Couture, 2008). It is a dynamic process affecting the concave side of the bank while depositing sediments on the opposite side. This erosion displaces people and create devastation work along the river course. Unlike other rivers of India, the rivers of North Bengal experiences massive bank erosion and it also causes human displacement and other problems. Maynaguri block is drained by two major rivers- the Teesta which is flowing along the western fringe and the other is the Jaldhaka flowing along the eastern part of the block. These two rivers cause massive bank erosion every year during the monsoon season. People suffer a lot due to the bank erosion of these two rivers. However, the present study first describes the global river bank erosion induced human displacement with a brief account of Indian example. After mentioning the bank erosion and displacement of people of Darjeeling, Jalpaiguri and Kochbihar districts the study proceeds to mention the vulnerable areas of Maynaguri block. An account of human displacement and other losses caused by river bank erosion of the Maynaguri block has been highlighted. The paper ends with some preventive measures.

Objectives

The present study embraces the following objectives-

- To identify the spatial distribution of bank erosion
- To present the displacement of people in Darjeeling, Jalpaiguri and Kochbihar districts as well as in the study area
- To suggest preventive measures

Literature Review

River bank erosion is a global phenomenon and causes displacement of people in different parts of the world. Since the problem is related to the socio-economic, cultural and political aspects many serious studies were carried out by different scholars. Mukherjee (2008) in his study of Malda and Murshidabad district showed that until now, more than 700 000 people in these two districts have been displaced due to Ganges bank erosion. There is a general tendency of River Ganges to shift towards the left bank above Farakka Barrage (Malda district), and towards the right bank below Farakka Barrage (Murshidabad and other districts). Rudra (2005) in his study showed that the river Ganges eroded 356 km2 of fertile land from the district of Murshidabad. He also mentioned that nearly 80 000 people in this district were displaced from 1988 to 1994. Halder & Das (2011) in their study mentioned the displacement of people due to flood erosion of river Ganges and its tributaries in Murshidabad district in 2000.

Showkat (2010) studied the effect of flood and associated land erosion and socio-economic aspects of Malda and Murshidabad area in details. The rate of river bank erosion is very high and frequent, which causes a huge amount of river bank cutting and population migration from the villages near the river every year (Iqbal 2010). Saha (2011) in his study mentioned the changes of course of the River Ganga in Malda district and its impact on Economy and society. Sarkar, et. al., (2011) in his study in the Sub-Himalayan North Bengal rivers puts forward the nature of bank erosion and displacement of people of the Teesta,Mahananda,Balason,Kaljani Lish,Ghish Diana Dima Jainti and Pagli rivers.

Database and Methodology

The present study is based on both primary and secondary data generated through author's field survey. In this study, the relevant secondary data were collected from the research-based books and the Disaster Management Section (Relief) of Maynaguri Block Development Office. The obtained data help the author immensely providing necessary information. The information on the human displacement of different continents due to hydrologic disasters was collected from the international organisations like IDMC and WPDS. For representing the displaced & affected people by river bank erosion in Darjeeling, Jalpaiguri, and Kochbihar districts from 2003 to 2015 the author has compiled various research-based data.

Data on river bank erosion induced displacement and other losses were collected from the Disaster Management Section (Relief) of Maynaguri Block Development Office from 1st August to 2nd August,2015. The author carried out a field survey along the Teesta erosion prone areas of Maynaguri block from 5th August to 8th August and identified the most vulnerable erosion prone areas from where human displacement took place. Finally, several protective measures like structural and non-structural measures have been mentioned. A literature review has also been done based on available books on river bank erosion.

However, all these collected primary and secondary information were later on systematically processed, arranged, tabulated and analyzed to have a clear view of the erosion induced human displacement scenario of the present study area. The information collected from primary and secondary sources have been verified with the field experience.

The Present Study Area

The present study area Maynaguri block lies in the western part of Jalpaiguri district in West Bengal. The block comprises 16-gram panchayats having a population of 3.29 lakh (Census 2011). The block is demarcated by river Teesta in the west and river Jaldhaka in the east. The northern fringe is bounded by Mal block while the southern part is touched by Koch Bihar district. The climate of the study area is characterized by hot and humid summer season. The area receives copious rainfall during the monsoon season i.e., from June to September.

Global River Bank Erosion Scenario

River bank erosion is a global ongoing natural process. The bank erosion process in several sections of the river network is influenced by the size of the channel, discharge, and flow strength (Florsheim et al., 2008). This bank erosion occurs mainly in meandering rivers. In meandering rivers, river-channel migration takes place through erosion of the cut bank and deposition on the point bar. However, all continents are more or less affected by the problem of river bank erosion. But its impact on population varies as the socio-economic conditions are different for different areas. Table-1 shows the percentage of continent-wise displaced population in the world due to all types of natural disasters in the years 2010 and 2011.The most affected region was Asia and least affected was Oceania. The displacement was mainly caused by hydrological disasters, which include floods, storms, and wet mass movements. Rest of the human displacement was due to geophysical (earthquake, volcanic eruptions, etc.), meteorological (storms), climatological (extreme temperature, etc.) and biological disasters (epidemics, etc.). But from the devastation point of view, the Mississippi-Missouri River System of USA, Ganges and Brahmaputra of India, Padma River, Jamuna River and Meghna River of Bangladesh and Mekong River of Myanmar, Laos, Thailand, Cambodia, Vietnam, Amazon River of Brazil, and River Nile of Egypt are the most vulnerable rivers of the world.

Table 1: Human Displacement in Different Continents (in percentage) from 2010 to 2011

Continents	Continent Percentage of Total Displaced Population in The World		Percentage of Displaced Population Relative to The Total World Population	
	Year 2010	Year 2011	Year 2010	Year 2011
Africa	4.01	3.91	0.0246	0.0084
Americas	19.34	6.76	0.1186	0.0145
Asia	76.43	89.12	0.4685	0.1906
Europe	0.15	0.12	0.0009	0.0003
Oceania	0.07	0.09	0.0004	0.0002

Sources: IDMC, 2011, 2012; WPDS, 2011.

Indian Scenario

North India

India is also a land of rivers like other Asian countries. Seven major rivers along with them numerous tributaries make up the river system of India. Floods, which are recurrent phenomena in India, cause severe bank erosion. Two rivers that are subject to severe bank erosion are River Ganges and Brahmaputra River (Sarma, 2013; Mili et al., 2013; Phukan et al., 2012). Various steps like bank embankment have been taken up for flood control in different countries but the erosion along the embankments and natural banks of the river systems is still a serious problem. The satellite image estimation shows that the total land loss per year due to erosion of Brahmaputra ranges from 72.5 to 80 km2/ year from 1997 to 2007 - 2008. Bank erosion has wiped out more than 2500 villages and 18 towns including sites of cultural heritage and tea gardens, affecting the lives of nearly 500 000 people.

The flood and erosion problems are serious in the lower Ganges region, particularly in West Bengal. The Ganges bank erosion in West Bengal is found to be severe in Malda and Murshidabad districts Land loss due to erosion in Malda is worse, too. More than 200 km2 of fertile land in Malda district have been completely wiped out until 2004. The displacement of of people due to erosion in these two districts is very acute.

Central India and Deccan Region

The important rivers in this region are Narmada, Tapi, Mahanadi, Godavari, Krishna and Cauvery. These rivers have mostly well-defined stable courses. They have adequate capacity within the natural banks to carry the flood discharge, except in the delta area. The lower reaches of the important rivers on the East Coast have been embanked, thus largely eliminating the flood and erosion problems.

Spatial Distribution of Bank Erosion in Darjeeling, Jalpaiguri and Kochbihar District & Maynaguri Block

Bank erosion is a common phenomenon in any uncontrolled channel of the deltaic tract. But in North Bengal this phenomenon is active in selected rivers. Bank erosion is prominent in Malda district caused by the river Ganges. Now it is well accepted that the intensity of the Ganges-bank erosion in West Bengal has increased after the construction of Farakka Barrage (Rudra, 2010). The Ganges in Malda district in the upstream of Farakka Barrage has been undergoing extensive erosion along the left bank even though it has been strongly protected. In Darjeeling, Jalpaiguri and Kochbihar districts also bank erosion is active. Table-2 shows the affected and displaced people and other losses due to bank erosion combined with flood from 2003 to 2010 in Darjeeling, Jalpaiguri, and Kochbihar districts.

Table 2: Displaced & Affected People by River Bank Erosion in Darjeeling, Jalpaiguri, and Kochbihar Districts from 2003 to 2015

Year	Affected Areas	Loss Scenario	Displaced & Affected people
2003	Jalpaiguri & Kochbihar district	170 m embankment of Irrigation & water dept. destroyed in Koch bihar 215 m long embankment washed away in Hasimara-Malbazar-Banarhat area 500 acre of tea garden land destroyed	31882
2004	Lish-Gish catchment areas of Jalpaiguri district	200m left embankment of Lish river destroyed	1000
2007	Jalpaiguri district	6500 homesteads destroyed and 4500 hectares of arable land destroyed	25000
2009	Darjeeling and Jalpaiguri district	Severe bank erosion along Balason river. Railway bridge over Balason near Matigara was destroyed	25000
2010	Jalpaiguri & Kochbihar district	Embankments in many places of Kochbihar and Falakata district were damaged	414
2015	Jalpaiguri district	3 km Pradhan Mantri Gramin Sarok nYojana(Road) near Domohani of Maynaguri block was severely affected Damage of embankment in Basusuba, Raipara & Senpara areas of Chadanga Gram panchayet of Mal block.	4000

Source: Sarkar et. al, (2011) & the field survey by the author (2015)

Maynaguri block is drained by several rivers. The block is mostly bounded by the river Teesta on its western sides. This river touches Domohani-I, Domohani-II, Barnesh, Padamati-I, Padamati-II and

Dharmapur gram panchayats. On the other hand, another wrathful river, the Jaldhaka originating from the Bitang lake area in Bhutan at an altitude of 4220 m touches Maynaguri block near Ramshai gram panchayat area. The Teesta and the Jaldhaka carry huge capacity of run-off during monsoon gushing down through their deep gorge with very high velocity and on reaching the flatter tracks over spill their banks resulting in severe bank erosion. These problematic rivers in this region, perennial in nature, cause occasional floods and the flood situation becomes worse during heavy rainfall when these rivers receive high rainfall in their catchment areas. The Jarda, another small river originating from Dooars forest area flows through Maynaguri block and receives the waters from neighbouring area in addition to its capacity and inundates its neighbouring localities with bank erosion, causing damage of dwelling houses of the indigent families who live on its both sides (Table 3).

Table 3: Major Bank Erosion Vulnerable Areas of Maynaguri Block

Name of The Gram Panchayets	Vulnerable Villages	River That Causes Bank Erosion
Domahani- I	Barmanpara, Chatrarpar	River Teesta
Padamoti - I	Dharparur char	River Teesta
Padamoti - II	Uttar Padamati char, Dakshin Padamati char	River Teesta
Dharmapur	Dharmapur char	River Teesta
Barnesh	Marichbari char	River Teesta
Amguri	Betgara colony	River Jaldhaka
Churabhander	Ratherhat char, Churabhander	River Jaldhaka
Ramsai	Dobbari & Khashi Morcha	River Jaldhaka

Source: Disaster Management Section (Relief) of Maynaguri block development office (2015).

Displacement of People and Other Losses Caused by Bank Erosion in Maynaguri Block

Impacts of river bank erosion are diverse: social, economic, health, education and sometimes political the first and foremost impact is social, i.e., homelessness due to land erosion which compels displacement of people. After displacement they suffer from economic crisis, namely loss of occupation and loss of property, and they are at the risk of poverty. The recent flood of Teesta and Jaldhaka river from 30th June to 2nd July, 2015 due to heavy rainfall and discharge brought a severe bank erosion. The eastern parts of river Teesta were severely damaged. River Jaldhaka also caused bank erosion. Table-4 shows the loss scenario of the present study area. About 1000 thousand people are now as yet living on the banks or nearby areas of Teesta and Jaldhaka river. The table reflects that the loss of land by Teesta bank erosion was 800 hectares while it was 300 hectares by river Jaldhaka. Total number of displaced people was 3000. Total number of damaged houses was 240.

Table 4: Loss Scenario of River bank Erosion of Maynaguri Block from 30th June to 2nd July, 2015

Name of The River	Affected Areas (Cultivated & Habited Areas)	Loss of Land (In Hectares)	Affected & Displaced Farmers	Loss of Houses
Teesta	Barman para, Daspara of Domohani; Chatrarpar, Basusuba of Mal block; Marichbari, Ulardabri of Barnish Sand bar areas of Dharmapur, Padamati –I Gram panchayet (Dhanir hat & Kamartari village)	800	3000	150
Jaldhaka	Betgara colony, Basilardanga, Bangkandi of Amguri	300		90
	Total	1100		

Source: Maynaguri Block Agriculture Office (2015)

Protective Measures

There are several common measures to reduce the bank erosion. These include the structural and non-structural measures.

Structural Measures

Riprap made of rocks and other materials, arranged in a way as to inhibit erosional processes on a river bank. This method is expensive and can experience failure, but has the ability to be used for large areas.

Windrows are the piling of erosion-resistant material on a river's bank, where if buried, they become known as trenches. When erosion persists an already determined location, these windrows and trenches are made to slide down with the bank in order to protect it from further occurrences of erosion.

Gabions are stacked, rectangular wire boxes filled with stones. They are useful on steep slopes when the water is too fast for the use of a riprap technique. They are expensive and labor-intensive, as well as require periodical inspection for damage and subsequent maintenance, though they have been seen to demonstrate positive performance.

The exact placement of soil cement may be different depending on the slope of the bank. In rivers with high wave action, a stair step pattern may be needed to dissipate the energy coming from the waves

Three main types of vegetation exist to prevent bank failure: Trees, shrubs, and grasses. Trees will provide for deep and dense root systems, increasing the stresses a river bank will accommodate. Shrubs are staked into the river bank in order to provide a protective covering against erosion, creating good plant coverage and soil stability.

In the erosion prone areas spur dams are very effective where meandering bends cause bank erosion.

Non- Structural Measures

Non-structural measures include the awareness of people about their living site. People should not build their homes on the raised sand bars and nearby bank areas because these sites are the most vulnerable to erosion. In the present study area, most of the people built their houses on the above site and so they have been victimized much. Hence people should be motivated not to build their houses on the sand bars and nearby bank areas. The role of Village level Disaster Management Committee should be active and effective for the distress people.

Finding

The findings of the present study show that Maynaguri block is vulnerable to river bank erosion. River Teesta and Jaldhaka are responsible for this erosion. These two-river cause severe bank erosion for which many people are displaced. Cultivated and habited areas are also lost due to this hazard. Displaced people become unemployed and homeless as they loss their habitation and agricultural lands. It is neither possible to stop bank erosion nor to completely eliminate its damages. Bank erosion occurred in the past and will continue to occur in future as well in this part of India. However, it is possible to minimize the severity of the impact and the damage potential. In this case mitigation measures like structural and non- structural measures have to be adopted in the present study area. Active and effective role of Village level Disaster Management Committee is essential for the distress people.

Conclusion

River bank erosion is a natural process of a stream. We cannot prevent it. Therefore, proper propaganda should be carried out by the respective gram panchayet offices to keep away human settlement from the erosion prone areas. Local Gram panchayet office can play an important role with the help of respective Block Development Office to build structural measures in the erosion prone areas.

References

Couture, S. (2008). 'River Dynamics and Erosion, Presented to Great Bay Siltation Commission, December 1, 2008', Conference paper. Online version (accessed 18 September 2014).

Florsheim, J. L., Mount, J. F. and Chin, A. (2008). 'Bank Erosion as a Desirable Attribute of Rivers', Bioscience, 58(6): 519–529, [DOI]. Living Reviews in Landscape Research http://www.livingreviews.org/lrlr-2014-3.

Halder,S & Das, B (2011). 'Disaster and Perception: A case study of 2000 flood scenario of Murshidabad district'. Published in editorial book 'Disaster Perception and Mitigation' (Sarkar et. al.). Department of Geography & Applied Geography: North Bengal University.

IDMC (2011). 'Displacement due to natural hazard-induced disasters: Global estimates for 2009 and 2010', Geneva (Internal Displacement Monitoring Centre, Norwegian Refugee Council). Online version (accessed 10 September 2015): http://www.internal-displacement.org/publications/2011/displacement-due-to-naturalhazard-induced-disasters-global-estimates-for-2009-and-2010. (Cited on page 14.).

IDMC (2012). 'Global estimates 2011: People displaced by natural hazard-induced disasters', Geneva (Internal Displacement Monitoring Centre, Norwegian Refugee Council). Online version (accessed 8September 2015): http://www.internal- displacement.org/publications/2012/

global-estimates-2011-peopledisplaced-by- natural-hazard-induced-disasters. (Cited on page 14.)

Iqbal, S. (2010). 'Flood and Erosion Induced Population Displacements: A Socioeconomic Case Study in the Gangetic Riverine Tract at Malda District, West Bengal, India', Journal of Human Ecology, 30(3): 201–211.

Mili, N., Acharjee, S. and Konwar, M. (2013). 'Impact of flood and river bank erosion on socioeconomy: A case study of Golaghat revenue circle of Golaghat district, Assam', International Journal of Geology, Earth & Environmental Sciences, 3(3): 180–185.

Mukherjee, P. (2008). 'CRY of Help for Erosion Victims', Business Standard, (February 19). URL (accessed 18 September 2014): http://www.business-standard.com/article/economy-policy/cry-of-help-for-erosionvictims.

Phukan, A., Goswami, R., Borah, D., Nath, A. and Mahanta, C. (2012). 'River Bank Erosion and Restoration in the Brahmaputra River in India', Clarion, 1(1): 1–7. URL (accessed 18 September 2014): http://clarion.ind.in/index.php/clarion/article/view/17. (Cited on pages 13 and 17.)

Rudra, K. (2005). 'The Encroaching Ganga and Social Conflicts: The Case of West Bengal, India', online resource, Gangaepedia (GRBMP). URL (accessed 18 September 2014): http://gangapedia.iitk.ac.in/?q=node/570.

Rudra K (1996a). 'Problems of river bank erosion along the Ganga in Murshidabad district of West Bengal'..Indian Journal of Geograhy & Environ 1:25- 32.

Saha, S (2011). 'Changes of course of the River Ganga in Malda district and its impact on Economy and society'. Published in editorial book 'Disaster Perception and Mitigation' (Sarkar et. al.). Department of Geography & Applied Geography: North Bengal University.

Sarkar et al. (2011).'Disaster Perception and Mitigation'. Department of Geography & Applied Geography: North Bengal University.

Sarma, D. (2013). 'Rural Risk Assessment due to Flooding and Riverbank Erosion in Majuli, Assam, India'. Master's thesis, University of Twente, University of Twente. URL (accessed 18 September 2014): http://www.itc.nl/library/papers_2013/msc/gfm/ sarma.pdf.

Uttar BangaSambad, a Vernacular Bengali daily, dated 09-09-2015/18-09-2015, Siliguri Publication.

15

Bankline Migration of Nonoi River and its Impact on Human Life: A Case Study of Sipajahr Development Block, Darrang District, Assam, India

Dr. Sikha Rani Nath

Assistant Teacher, Assam Govt. Service, Guwahati, Assam, India

Abstract

The study area Sipajhar Development Block is prone to flood and bank erosion every year because of its topographical and hydrological characteristics. The river Nonoi passing through this region create high magnitude flood due to its high volume of water in every year. Bankline migration due to erosion is a common feature in the River Nonoi and it is the most devastating river of the study area which plays great impact upon physical and socio-economic environment. For the study both primary and secondary data are collected and the data are analysed by using some statistical techniques. For scientific and inclusive analysis a multi temporal database have been developed based on Survey of India toposheets (1967) and satellite imageries (LISS-III).Remote sensing and GIS technologies are used to find out the migration of the river bank line over a period of 18 years from the year 1999 to 2017. It is found from the study that the bankline migration is caused due to bank erosion and in both sides of the river bank erosion is more pronounced compared to sedimentation. The maximum shifting is found about 1.20 km in the village Phuhuratali where the river bankline is shifting in westward direction.

Keywords: Nonoi River, Bankline Migration, Bank Erosion, Flood, GIS.

Introduction

Rivers have been considered as the lifeline of living beings and history recalls that the great human civilizations were flourished in the river bank. As the river flows, the dynamics of the exchange between its water and sediment load and the geology and hydrology of the landscape creates a complex network of processes, resulting in a given physical form of the river system. Bankline migration is one of the important fluvio- geomorphological processes through which a river tries to attain equilibrium condition. Channel changes as result of combined process of erosion, transportation and deposition. Such phenomena involve the removal of material from one side of the basin to deposit the same in another side and thus there occur channel shifting and channel metamorphosis. The monsoon dominated rivers are characterized by high variability of discharge and sediment yield which makes the river unstable. These unstable channels are prone to floods of high to medium intensities as they are small enough as compared to the total supply of waters along the channel. Such situation is associated with rapid degradation and aggradation processes and changes the morphology of the river basin.

The River Nonoi is one of the north bank tributaries of the River Brahmaputra, Assam and it is the biggest river of Sipajhar Development Block flowing in the central western part of the area. The river has its origin in the Tongsa province of Bhutan Hills at an elevation of 1220 meter above mean sea level. Initially it was known as Degauge and as time passes the river changed its older course and formed a new stream which was later known as Nonoi. Till date the river has been shifting its course towards west direction from its original channel. The existence of the river Bengnoi is the indication of previous course of the river Nonoi. It has intruded and divided in to several channels, ox bow lakes and swamps on its abandoned course. The length of the river flowing across the district is 53 km and through the studied area is approximately 48 kms. Due to heavy downpour during summer months the river carries huge volume of water subject to serious flood in the entire area causing extensive damage to crops and people, soil and bank erosion on both banks. Nonoi exhibits as the most flooding and erosive river among the different rivers of Darrang district. The River is characterised by heavy flood during the rainy season, enormous volume of water and sediment load, bankline migration and channel erosion which further causes continuous changes of channel morphology of the River.So Floods of very high magnitude with high frequency can be considered as a contributing factor to bank erosion and channel widening which cause channel migration of the river Nonoi in the studied area . Abrupt changes of flow pattern are caused due to heavy flooding by Nonoi river in the study area. The processes of bank erosion are important in the development and evolution of different channel form, while the migration of river channels across their floodplains involves a combination of bank erosion and deposition. The record of flood history together with breaches of embankments indicates that large floods have occurred every year caused bank erosion, aggradation and channel migration in the studied area.

The magnitude and frequency of flows and the probability of both the low and high flows in Nonoi River have a great significance upon the existing landscapes. The problems of bankline migration along with flood and river bank erosion are the dynamic fluvio-geomorphic processes which have adverse effect on society, economy and environment as a whole. The potentiality of damages is closely related to the nature of occupancy, stage of socio-economic development and various physical parameters flood and river bank erosion.

Study Area

The study area is confined to Sipajhar development block in Darrang District which extends latitudinally from 26° 12′ 42″ to 26° 32′ 25″ and longitudinally from 91° 44′ 49″ to 91° 57′ 18″. The area is located in the middle part of the Brahmaputra valley of and is surrounded by river Saktola in the east, Barnadi in the west, Mangaldai-Mazicuchi road in the north and the mighty Brahmaputra in south.

The study area presently comprises 178 villages in twenty five GP with a population of 2,19,824 people (2011 census) and area covering of 405.38 square kilometers, the district is having a density of 589 persons per sq km. Different types of ethnic groups are living in this region and the socio economic condition of the people varies from place to place, where the average literacy rate is 67.53%.The main occupation of the people in the district is agriculture (70-80% people engaged) . There are seven development blocks in the District among which Sipajhar Development Block is worst flood affected area of the District. From the field study and by examining the relief and contour, it is clear that the Sipajhar Development Block is characterized by diverse topography. The landform is gently sloping towards the south and the middle part of the area is relatively lower than the eastern and western part. Therefore, the area signifies different fluvio geomorphic processes such as bankline migration, bank erosion, river congestion, flood etc. and the morphology of the area are shaped by the combined effect of fluvial processes of the river along with human interference. The location map of the study area is given in the figure 1.

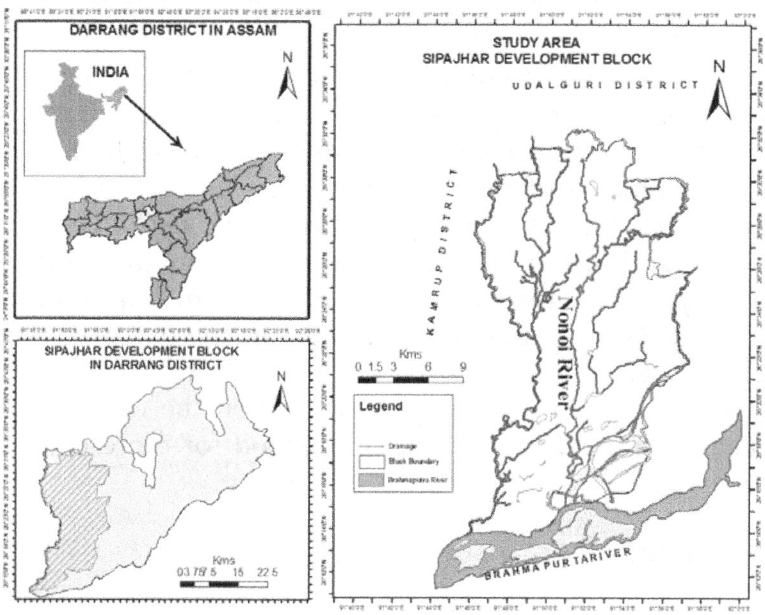

Figure 1: Location Map of The Study Area

Objectives of the Study

The present study is aimed at the following objectives:

1) To study the bankline migration pattern of Nonoi River in the study area on spatio-temporal basis.
2) To examine the causes of bank line migration and erosion on the study area.
3) To study the impacts of bankline migration on human lives.

Materials and Methods

For collection of various types of data both empirical and field survey is done. There are two principal sources of data viz, primary data are collected from direct field observation and survey and secondary data comprising books, journals, magazines, newspapers, articles, maps, satellite images etc. SOI topo sheets of 1967 at 1:50000 scale with No. 78 N/14, 78 N/15 and 78 N/16 were used to delineate drainage, water bodies, wetlands, sandbar etc. of the study area. For extraction of spatial as well as temporal changes of bankline migration of Nonoi River, IRS ID and IRS P6 LISS-III imageries dated of 19/12/99 and 13/11/17 having a spatial resolution of 23.5 m were used. Data pertaining to feed up different objectives are collected from different respective sites.

After selecting the study area different base layers such as Block boundary, village boundary, district boundary, drainage, water bodies etc. have been prepared using GIS software (Arc GIS). The first step is the georeferencing of the imageries using GIS software and then digitized the banklines for both years from the images. The banklines are then overlaid and it gives the total channel migration from 1999 to 2017 Extensive fieldworks are done in order to collect all relevant and necessary data and to gather information and experience through keen observation of ground reality. The whole process involves sorting, tabulation and classification of the data. Then the dataset has been manipulated using suitable statistical method for representation and analysis.

Result and Discussion

Like other alluvial rivers the Nonoi River also displays a tendency to migrate its bankline. In the studied area the river exhibits migration in some reaches where shifting of bankline is likely due to erosional and depositional processes. For determining the bank erosion two years of satellite imageries over a span of 18 years (1999 to 2017) have been considered. The river Nonoi is carrying high volume of water and sediment from the Bhutan hills and depositing in the lower reaches which tend to bankline shift. In the middle reaches due to shifting activity the River Nonoi captured the sub tributary Bengnoi as a result of which it become a dead river in the area. In Nagaon village the river has been shifting its course eastward up to a distant of half kilometer from its original path. In the village Dhokapara the river shifts its bankline up to a distance of 1 km and meander scars are formed due to progressive migration of the river bankline. In kuruwa village, the southernmost village of the block the River Nonoi is again shifting towards east direction leaving its mean path. The maximum shifting is found in the village Phuhurtali about 1.20 km where the river is shifting in westward direction. The subtributary Dimila, kalapani and Ghagra was previously following the dendritic pattern but now it has changed its direction and used to flow parallel with the Nonoi River.

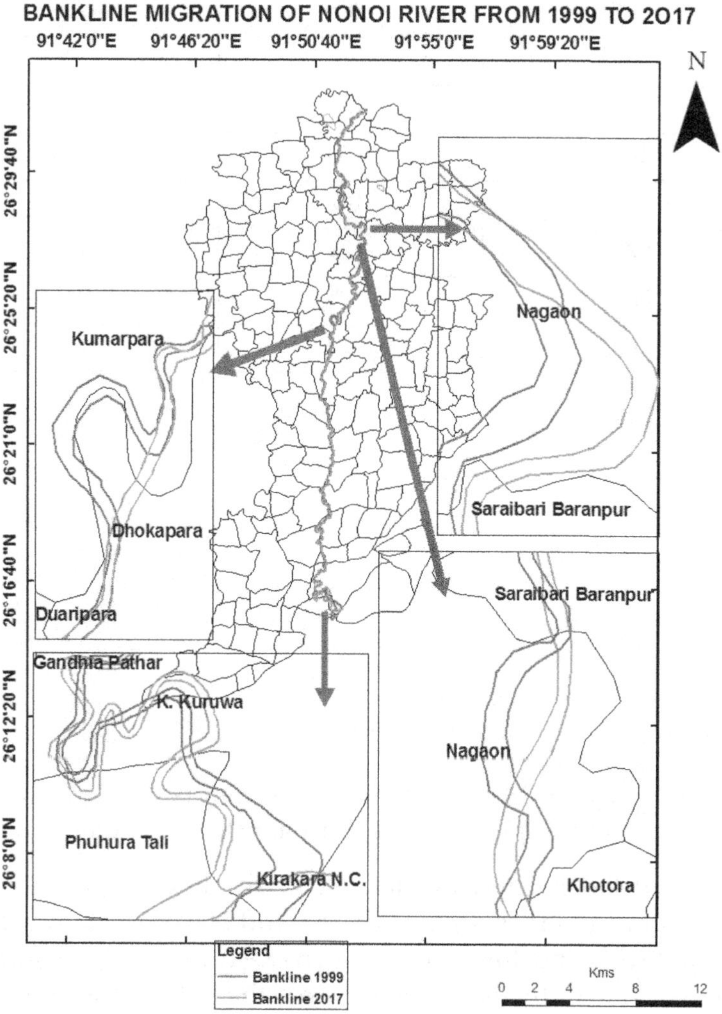

Figure 2: Bankline migration of Nonoi River from 1999 to 2017

Both the banklines of the river are digitized for two different years and after overlaying the layers it is found that the river Nonoi is shifting its courses and erosion and deposition occurs both in the right and left bank of the River. The shifting of the river from 1999 to 2017 along both the banks was measured in 9 cross sections and presented in the table 1.

Table 1: Changing Course of Nonoi River from 1999 to 2017 in meter.

Sections	Left Bank	Right Bank
AA″	-216	-240
BB″	-205	-223
CC″	467	489
DD″	-133	-140
EE″	-130	-125
FF″	-147	-151
GG″	-172	-160
HH″	-195	-245
II″	-322	-582

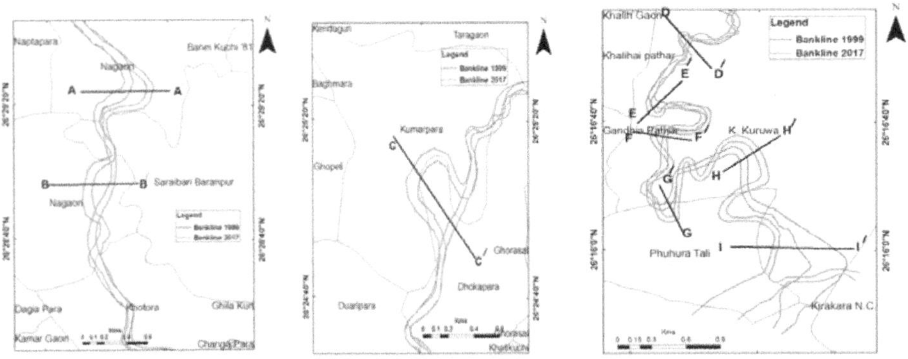

Figure 3: Changing Course of Nonoi River from 1999 to 2017 in Sipajhar Development Block, Darrang District, India

It is seen that highest erosion along the both right bank and left bank are found in cross section II″ (82 and 322 meter) and the rate of deposition are found highest in the cross-section CC″. Minus sign (-) indicates shifting of the bankline towards the left bank and plus (+) sign indicates the shifting of the bankline towards right bank from the year 1999 to 2017. In the fig 4 it is seen that erosion is more dominant along the right bank of the river which indicate shifting of the bankline towards western side. In the cross-section C, the amount of sedimentation indicates the shifting of the left bank towards eastern side as in this section the river is found to shift about half km from its original path.

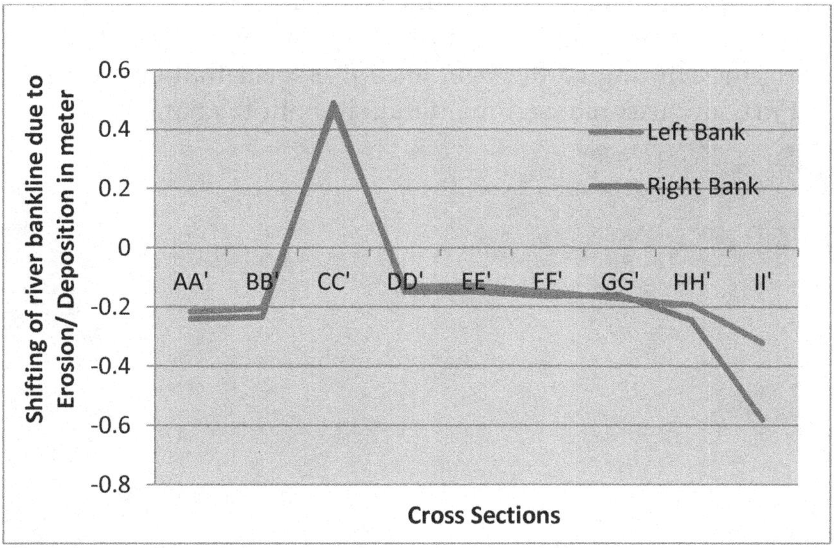

Figure 4: Changing Course of Nonoi River from 1999 to 2017 in Sipajhar Development Block, Darrang District, India

The River Nonoi in the studied area is found to be more disposed to river bank erosion and sediment deposition. Erosion may be caused either by undercutting the upper bank materials by river during high flood times or by over steepening of bank materials due to migration of thalweg closer to the bank during the falling stages [6]. Among the physical impact land use land cover change is one of the most significant impact of flood and riverbank migration. The wetlands, ponds and the rivers are studded with silt and sand deposition. The shifting of channel and bank erosion of the river Nonoi caused damage to the land and people of this area. The extent of bank erosion and sand deposition along the Nonoi had become erratic and unstable causing serious damage as the widths of both the banklines are increasing continuously. It creates displacement of settlement, residential area loss, productive and agricultural land loss etc. and increases the number of land loss labourers. Due to the uncontrollable nature of bank lines of the river anti-erosion projects adopted for the protection of land and people have become a failure at the villages of Khatara, Dipila, Byaspara and Khandajan, etc. There was rapid bank erosion on both the heads of the bridges on Nonoi River due to a great flood in the Patharighat Circle in the year 2014. During 1999-2017 there was a total loss of 3600 hectares of land in occupied villages. In the village Nagaon the river erodes its left bank accounting 80.71 sq.km and 74.81 sq. km. on its right bank. From analysing the shifting of the banklines it is clear that erosion is more evident in comparison to sedimentation in both the banks.

Plate 1: Bank Erosion by Nonoi River in The Study Area

Plate 2: Anti Erosion Measures by Government at Nonoi River Bank

From the official report it is observed that Nonoi River is the most harmful river as the loss and damage of embankment caused by flood ranks first among all the river of the district. Nonoi is a major river is characterized by exceedingly large and variable flow, enormous volume of sediment transport and sedimentation, changes in channel configuration and bank erosion. The important factor determining the migration of the river bank line in the studied area is the presence of fine and silt soils which are very less resistance. In the year 2004 during the months of May, June and July River Nonoi had attained record of flood level of 54.77 m. Again, in the year 2008 and 2014 measured at NH 15 crossing against its danger level of 52.74 meter caused severe flood in the entire area. Floods, bank erosion, bank line migration and sedimentation have created adverse effect on agriculture. Agriculture is the main occupation of the people in this area but floods in every year cause destructions of the paddy and other rabi crops which affect the economy of the area. Agricultural land loss due to erosion and siltation is a common phenomenon in this region. The major loss due to flood is announced by crop damage and crop damage is serious for most of the people who do not have surplus. In 2016 a total of 2,321 hectare of cropped area in the Block are affected by flood and bank erosion. Bodo and Ahu are the alternate cultivation in this area to attain sustainability in crop production. That is flood associated with Bankline migration may not only cause immense financial losses but also causes deformation of landscape, environmental pollution, inhibition of economic activity, destruction of economic infrastructure and property and innumerable social loss. The villages which are very badly affected by bank line migration and erosion by the Nonoi river are Kumarpara, Dhokapara, Ghorasal, Duaripara, Gopalpur, Nagaon, Dhekipara, Punia, Badiasicha, Narikali, Singimari, Borabari, Kurua, Fahurahati, etc. The numbers of villages affected by floods of River Nonoi have been increasing in recent years. The frequently occurring Channel shifting, floods and bank erosion in the active floodplain areas have been executing outmigration of indigenous people who are unable to adjust with this situation. Various flood protection and anti-erosion measures are implemented by government to reduce the damages of erosion and migration by river. Constructions of structural measures such as embankment, dyke, sluice gate etc. are somehow efficient in controlling and reducing flood and erosion in some areas. But failure of these measures brings intolerable damages to the lives of the people as breaching of embankment is common in every flood time due to leakage, seepage, piping, erosion etc.

Conclusion

Floods associated with bank line migration and bank Erosion are the natural occurrence which have detrimental impacts upon the physical and socio-economic environment. For analysing the changing pattern of the banklines the imageries for different years 1999 and 2017 are digitized and overlaid to one another and it is found that due to both erosion and deposition the banklines are migrating from its original path. The river bankline of the Nonoi River is migrating towards western direction and the maximum shifting is found in the village Phuhuratali. From the field study and researcher's knowledge it is distinct that floods in this area arises 3 to 4 times in a year and it is sometimes so hazardous that the whole of the environment, economy and society totally disrupted. People are adjusting with flood and erosion to certain extent by accepting the losses as the traditional measures are sometimes withstand the flood hazard. Due to the increasing nature of magnitude and frequency of these natural hazards it is very essential for the hydrologists, planners and engineers to know accurately hydrological parameters for implementing various development plans. Deforestation along the river banks should be checked and afforestation should be encouraged under various programs. Frequency analysis of water flow is now considered as the most commonly used and sophisticated technique for the geomorphologists because of its simplicity and significance.

References

Baishya, S.J., Sahariah, D. (2015). A Study of Bank Erosion and Bankline Migration of the Baralia River, Assam Using Remote Sensing and GIS. *International Journal of Current Research, vol11*, 73-80.

Bhattacharjee, N. (2008). *Flood and Bank Erosion Problems in Darrang District, Assam: A fluvio Geomorphological Study.* an Uunpublished Ph.D Thesis, Gauhati University.

Bordaloi, N. (1986). Effect of the Brahmaputra Channel Migration on Flood Plain Occupance: A Case Study of the Palashbari-Garimari Tract in South Kamrup. *A M-Phil Dissertation , Gauhati University. Guwahati, Assam.*, 51.

Desai A.J.,Naik S.D., Shah R.D. (2012). Study on the channel migration pattern of Jia-Bhareli river, Puthimari and Pagladiya tributaries of the Brahmaputra river using REmote Sensing Technology. *htttp://www.isprs.org/proceedings /XXXVI/part4/pages from RS-GEO14.pdf.*

Gogoi, C., Goswami, D.C. (2013). A study of bank erosion and bankline migration pattern of the Subansiri River in Assam using Remote Seneing and GIS Technology. *The International Journal of Engineering and Science,vol.* 2, 1-6.

Goswami, D. (2002). *Channel pattern, Sediment Transport and Bed regime of the Brahmaputra River, Assam.* . (S. T. Thakur, Ed.) New Delhi: In: Recent Advances in Geomorphology, Quaternary Geology and Environmental Geosciences: Indian Case Studies. Manisha Publication,.

Sarkar, A., Garg, R.D.,Sharma, N. (2012). RS-GIS based assessment of river dynamics of Brahmaputra Riverin India. *Journal of Water resource and Protection,vol.4,* 63-72.

Schumn, S.A., Lichty R.W. (1963). Channel widening and Floodpalin construction along Cimarron Riverin southwestern kansas. *US Geological Survey.Prof Paper, 352-D.,* 71-88.

16

A Study on Arsenic Contamination of Ground Water in Lalgola Block under Murshidabad District, West Bengal

Ananya Chowdhury

Assistant Teacher, Swarupganj Jr. High School, Malda, West Bengal, India

Abstract

In the present age, the population is a terrible problem. Pollution is leading to the destruction of the environment. Arsenic pollution of water is a terrible pollution. The effect of this pollution are gradual but deadly. Water is life. Without water, human existence is endangered and because that water is polluted, mankind is in danger today. The groundwater of Lalgola block (study area) contains a large amount of arsenic. The inhabitants of the area depend on the water of the deep tube well. As it is a rural agriculture depended area, ground water is extracted for cultivation and irrigation. Resident of this region have been drinking arsenic- contaminated water for many years. Because of this they are suffering from various disease like black foot disease, arseniccosis, Cancer etc. many people have died of Cancer. The Government and NGOs are working together. The government has install some arsenic free tube-wells but not enough. People's awareness is low. For this, role of the Government and the educated people is more necessary. Awareness education on arsenic contamination is needed. That education should be from primary level. Extensive application of arsenic removal plant is required.

Keywords: Arsenic, Ground Water, Contamination, M.C.L, Blackfoot Disease, ARP, Cancer.

Introduction

Arsenic is a chemical element with the symbol as an atomic number 33. Arsenic occurs in many minerals usually in combination with Sulphur and metals, but also as a pure elemental crystal. Arsenic is a metalloid. It has various allotropes, but only the grey form which has a metallic appearance is important to industry .arsenic is widely found in Earth crust in two chemical forms inorganic and organic. inorganic arsenic is more toxic than organic arsenic .among inorganic arsenic trivalent from (arsenite) is more toxic than pentavalent from (arsenate) presence of arsenic in groundwater in the intermediate aquifers(20-80meters below ground level) may be anthropogenic or geogenic in nature. IN West Bengal it is geogenic. Inorganic arsenic of geological origin is found in groundwater used as drinking water in several parts of the world. Arsenic contamination of groundwater is a natural occurring high concentration of arsenic in deeper levels of groundwater, which became a high-profile problem in recent years due to the use of deep tube wells for water supply in the Ganges delta, causing serious arsenic poisoning to large number of people. It 2007 study found that over 137 million people in more than 70 countries are probably affected by the Arsenic poisoning of drinking water. Arsenic contamination of groundwater is found in many countries throughout the world including the USA. Arsenic is a carcinogen which causes many cancers including skin, lung, and bladder as well as cardiovascular disease outcomes included several types of cancer, diseases of the circulatory and respiratory system, Diabetes mellitus, kidney and liver disease. Elevated mortality rates were observed for all diseases of the circulatory system. [Some research concludes that even of the lower concentrations there is still a risk of arsenic contamination leading to major cause of death.]

Objectives

The objectives of the present paper are:

1) To enquiry the sources of arsenic in groundwater particularly Murshidabad district.

2) To study the arsenic affected blocks in Murshidabad district particular lalgola block.
3) To study the different affected people by arsenic contamination.
4) To study the proper mitigation policy for arsenic affected areas in Murshidabad district.

Methods and Approaches

The present work is very much dependent on intensive fieldwork. Data have been collected from personal contacts, discussions with the executives as well through questionnaire survey and various geographical records. And different published and unpublished sources. Photographs have been used to supplement the data the approaches have been qualitative and quantitative.

Discussion

Groundwater being usually easy and inexpensive in abstraction and safer is used as main source of drinking water in West Bengal during early part of 80s of some cases of people suffering from arsenic dermatitis in the district of 24th Parganas south, 24 Parganas North, Nadia, Murshidabad and Bardhhaman were detected through epidemiological studies conducted by School of tropical medicine, Kolkata and all India institute inch if hygiene and public health. In order to examine and investigate the matter, government of West Bengal immediately constituted a working group in December 1983. it was observed by the group that the sample of groundwater showed presence of arsenic beyond M.C.L (maximum containment level) of 0.05 mg/l. A problem of this magnitude was not encountered anywhere in national or international level.

Case Study of Lalgola Block

The study area that is lalgola block, the Panchayat of lalgola block are lalgola Paikpara, Ramchandrapur, Manik Chowk, Bahadurpur, Dewansarai, Nasirpur, Kalmegha, Airmary, Jasaitala, moiya are more affected by arsenic. There is a table regarding safe and unsafe tube wells (up to August 2011) of lalgola block.

From Table 1 shows that there are many unsafe tube wells in Lalgola Block, In Lalgola Panchayat, there were 213 unsafe tube wells. In Airmary panchyat, there were 60 unsafe tube wells. The Lalgola panchyat is highly affected area.

Table 1: Panchayat-wise Status of Tubewells in Lalgola Block

	Up to The Month (Up to August-2011)		
Panchyat	**Safe**	**Unsafe**	**Total**
Lalgola	191	213	404
Paikpara	220	94	314
Ram.Ch. Pur	211	146	357
Moiya	102	146	248
Manikchak	256	49	305
Bahadurpur	182	70	252
Bilboracopra	74	197	271
Kalmegha	188	116	304
Jasaitala	230	179	409
Dewansarai	129	68	197
Nosipur	142	34	176
Airmary	270	60	330
Total	2195	1372	3567

Source: Water Sample Testing Laboratory of Lalgola

Occurrence in Murshidabad District

1) Raninagar- I, 2) Raninagar-2, 3) Domkal, 4) Nawda, 5) Jalangi , 6) Hariharpara, 7) Beldanga-1, 8) Suti-I , 9) Suti-II, 10) Bhagawangola-I, 11) Bhagawangola-II, 12) Berhampore, 13) Raghunathganj-II, 14) Murshidabad-Jiaganj, 15) Farakka, 16) Samserganj, 17) Lalgola, 18) Beldanga-II, 19) Raghunathganj-I.

The Map-1 Shows that Suti-I, Suti-II, Raghunathganj-II, Lalgola, Bhagwangola -I, II, Rejinagar-I, II, Jalangi, Domkal, Hariharpara, Nowda, Beldanga,-I blocks were severely affected. Patients were found in all the above mentioned blocks except Raghinathganj-I,II,

Sagardighi,Murshidabad- Jiaganj, Berhampore, Beldanga-II were moderately affected blocks. Nabagram, Khargram, Kandi, Bharatpur-I, Burwan were mildly affected blocks, Bharatpur-II block is safe.

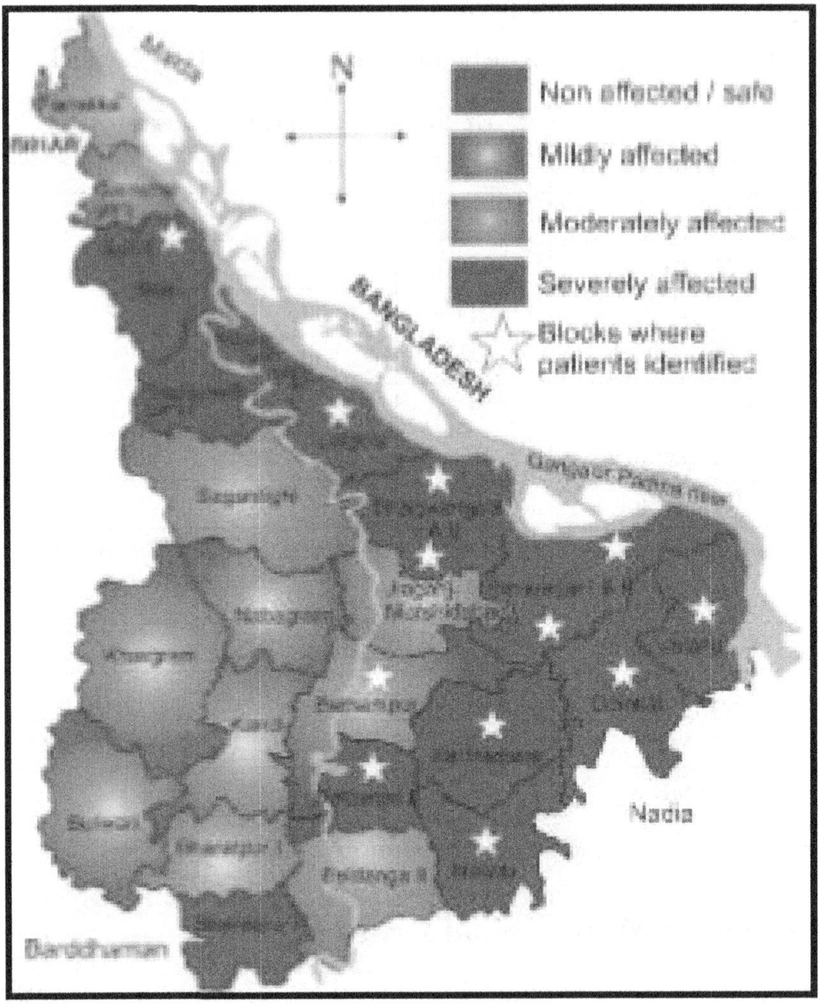

Map 1: Arsenic Affected Areas
Source: Google Image

From Table-2, we learn that laboratories have tested public and private water sources. Many of the tests performed in October showed arsenic- affected conditions, throughout the district. Another issue is that Public tube wells are more contaminated with arsenic than private tube wells.

Table 2: Water Sample Testing Report Murshidabad District

Sl. No.	Name of The Water Testing Laboratories	Name of The Block	No. Of Testing During the Reporting Month (October-2011)			No. Of Arsenic Affected Tube Wells (During the Month)		
			Public	Private	Total	Public	Private	Total
1	Sevavrata	Berhampore	190	16	206	71	3	74
		Beldanga-I	0	0	0	0	0	0
		Beldanga -II	90	0	90	23	0	23
		Bhagwangola-II	40	0	40	22	0	22
		other block	0	5	5	0	0	0
		Total	320	21	341	116	3	119
2	R.N.Club	Raninagar-I	80	3	83	32	2	34
		Raninagar-II	100	11	111	18	2	20
		other block	0	6	6	0	0	0
		Total	180	20	200	50	4	54
3	Amtala Village Organisation Society	Hariharpara	10	0	10	7	0	7
		Nowda	110	0	110	63	0	63
		Total	120	0	120	70	0	70
4	Farakka Block Bidi Shramik Kalyan Society	Farakka	0	0	0	0	0	0
		Samsherganj	0	0	0	0	0	0
		Suti-II	0	0	0	0	0	0
		Total	0	0	0	0	0	0

Sl. No.	Name of The Water Testing Laboratories	Name of The Block	No. Of Testing During the Reporting Month (October-2011)			No. Of Arsenic Affected Tube Wells (During the Month)		
			Public	Private	Total	Public	Private	Total
5	MIT	Bhagwangola-I	0	0	0	0	0	0
		Murshidabad-jiaganj	0	0	0	0	0	0
		other block	0	0	0	0	0	0
		Total	0	0	0	0	0	0
6	W.B.V.H.A	Lalgola	90	0	90	0	0	0
		Raghunathganj-II	93	0	93	0	0	0
		other block	0	0	0	0	0	0
		Total	183	0	183	0	0	0
7	Harpur Dr. Ambedkar Jana Seva Mission	Kandi	40	0	40	0	0	0
		Burwan	90	0	90	0	0	0
		Bharatpur-I	50	0	50	0	0	0
		Bharatpur-II	50	0	50	0	0	0
		Other block	0	0	0	0	0	0
		Total	230	0	230	0	0	0
8	PHED Jangipur	Raghunathganj-I	20	0	20	1	0	1
		Sagardighi	70	0	70	0	0	0
		Suti-I	40	0	40	1	0	1
		Total	130	0	130	2	0	2

Sl. No.	Name of The Water Testing Laboratories	Name of The Block	No. Of Testing During the Reporting Month (October-2011)			No. Of Arsenic Affected Tube Wells (During the Month)		
			Public	Private	Total	Public	Private	Total
9	PHED Domkal	Domkal	0	0	0	0	0	0
		Jalangi	0	0	0	0	0	0
		Total	0	0	0	0	0	0
10	PHED Mahyampore WTP	Beldanga-I	123	0	123	64	0	64
		Total	123	0	123	64	0	64
11	PHED Berhampore	Khargram	30	0	30	0	0	0
		Nabagram	47	0	47	0	0	0
		Other block	16	0	16	0	0	0
		Total	93	0	93	0	0	0

Source: Murshidabad Zilla Parishad.

Table 3: Duration of Time to Inhale (water)

5th Year	10th Year	15th Year	20th Year
2	8	10	10

Source: Surveyed and Compiled by Researcher

Table 3: Shows the duration of time to inhale arsenic contaminated water. Some families have been surveyed for how many years their water has been inhaled, (Table-3) meanwhile, it is seen that 10 families have been drinking contaminated water for 15 years, 8 families for 10 years and 2 families for 5 years, Families who have been drinking contaminated water for more than 5 years have been affected with arsenic-borne diseases.

Table 4: No. of Person in the Family and Affected Person

	Male	Female
Total Family Member	80	60
Affected Person	15	25

Source: Surveyed and Compiled by Researcher.

Table-4 shows that 15 male and 25 female members were affected. Female member were more affected than male member.

Table 5: Disease Category

1st Stage	2nd Stage	3rd Stage
10	25	5

Source: Surveyed and Compiled by Researcher.

In the survey area there are 10 people in the 1st stage of disease category 25 people in the 2nd stage and 5 people in the 3rd stage. Some people have died of cancer. (Table 5)

Effect on Human Health

1) Soluble inorganic arsenic is actually toxic and ingestion of large doses lead to gastrointestinal symptoms, disturbance of cardio-vascular and nervous system function and eventually death. In

survivors, bone narrow depression, hemolysis, hepatomegaly, melanosis, polyneuropathy and encephalopathy May be observed.

2) Long-term exposure to arsenic in drinking water is causally related to increased risks of cancer in the Skin lungs, bladder and kidney as well as other is skin changes a hyper keratosis and pigmentation changes. This effects have been demonstrated in many studies using different study designs. Exposure - response relationships and high risk have been observed for each of these end- points. The effects have been most thoroughly studied in Taiwan but there is considerable evidence from studies on population others countries as well. Increased risk of lung and bladder cancer and of arsenic associated skin lesions have been reported to be associated with ingestion of drink- water at concentrations 50 mg arsenic/liter.

3) Occupational exposure to arsenic, primarily by inhalation, is casually associated with long cancer. Exposure response relationship and highly risk have been observed. Increase risk have been observed at cumulative exposure levels 75 mg (mg\m3), year (e.g. for 15 years of exposure to a workroom air concentration of 50 mg\ m3). Tobacco smoking has been investigated in two of the main smelter cohorts and was not found to be the cause of the increased lung cancer risk attributed to arsenic; however was found to be interactive with arsenic increasing the lung cancer risk.

4) Even with some negative findings, the overall weight of evidence indicates that arsenic can cause lactogenic damage in different cell types with different end points in exposed individual and in cancer patients or point mutations, the results are largely negative.

5) Chronic arsenic exposure in Taiwan has been shown to cause Blackfoot disease (BFD), s severe form peripheral vascular disease (PVD) which leads to gangrenous changes, however, there is good evidence from studies in several countries that arsenic exposure causes other forms of PVD.

6) The evidence is strongest for Hypertension and cardiovascular disease, suggestive for diabetes and reproductive effect and

weak for cerebrovascular disease, suggestive for diabetes and reproductive effects and weak for cerebrovascular disease, long-term neurological effects, and cancer at sites other than lung, bladder kidney and skin.

Toxicity

Toxicity: acute and chronic

Acute: Ingestion

Serve vomiting, diarrhea, muscular cramps, facial edema and cardiac abnormalities, fatal ingested dose: 70-180 mg of arsenic (III).

Chronic (Arsenicosis): Repeated ingestion (water), inhalation (dust/fumes)

Systematic Manifestation: Multi-System Disease

- Chronic weakness, easy Fatigability, lack of mitivation, anorexia, wt and hair loss, mostly affected respiratory system.
- Respiratory system: chronic cough. Elevation of symptoms like cough with expectoration, resp distress, wheezing and ronchi in patients with ch. arsenicosis. Br. Carcinoma (late Stages)
- GI system: dyspepsia, nausea, anorexia, Ch diarrhea and vague abdominal pain.
- Hepatic system: Hepatomegaly. Non-cirrhotic portal fibrosis with/without portal hypertension. Cirrhosis.
- Renal System: Bladder and kidney cancer.

Arsenicosis Carcinogenic Effect

- Carcinogenic effect: Takes years to develop.
- Skin - Bowen's disease, basal cell carcinoma-trunk and aquamous cell carcinoma in the extremities.
- Resp Sys- Lung carcinoma.
- Genitourinary sys- bladder and kidney cancer.
- Liver-Hepatocellular carcinoma and hepatic angiosarcoma.

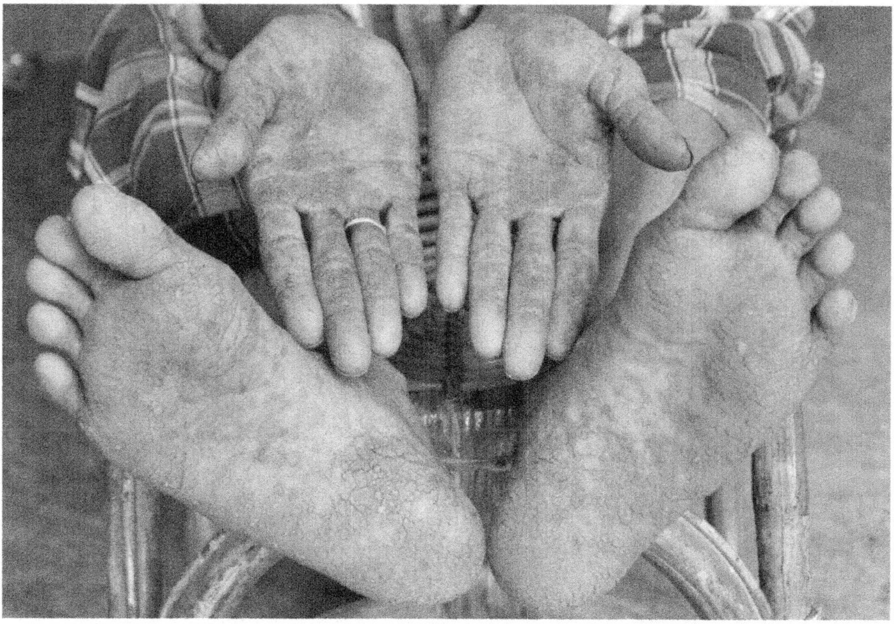

Plate 1: Effect of Arsenic on Hand and Leg

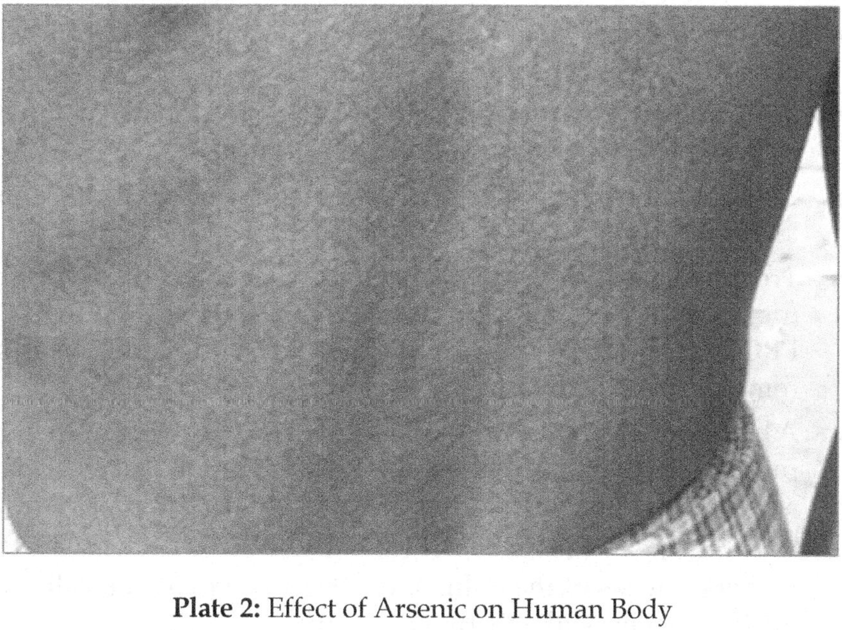

Plate 2: Effect of Arsenic on Human Body

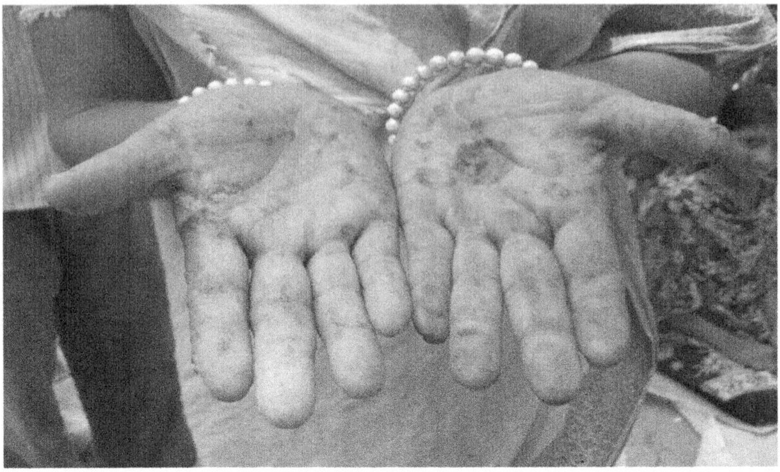

Plate 3: Arsenic Effect in Rural Areas
Source: Google Image

Management

- To stop drinking arsenic containmented water and to drink arsenic free water.
- To use surface water from reserved tanks ponds or from wells after proper disinfection.
- To get the personal or public tube wells tested for arsenic content even if these have been newly sunk.
- Case identification could be done at the Primary Health Centre level.
- Treatment with the chelating agents like DMSA (demercapto-succinic acid) or DMPS (2, 3 DIMERCAPTO-1-PROPANESULFONATE) 1 100mg 4 times /day for 7 days still remains questionable.
- A high protein diet form animal sources or Vegetable sources like pulses, soybeans, weight etc intake of plenty of leafy vegetable and fruits with plenty of vitamins and antioxidants (VIT A E and C) can help to improve clinical symptoms.
- Hyperkeratosis of the Palm and sole can be treated locally with 5-10% salicylic acid and 10-20% of urea.

- The systemic clinical manifestation complications could be treated symptomatically at sub divisional, district and tertiary hospitals.
- Health education, community awareness, community participation and mobilization to deal with the social problem of Arsenicosis.

Mitigation Policy

- Piped water supply- based on rivers or big diameter deep tube wells.
- Deep tube Wells- to tap water from safe aquifers, third or deeper layers beyond 100 meters.
- Arsenic removal plants attached to tube wells.
- Horizontal roughing filter (H R F) and slow sand filter (SSF) attached to ponds horizontal movement of water through graded gravel followed by sand and a disinfection process at end.
- Domestic filter-affordable and user-friendly treats water in two stages. at first iron is removed by a special type of candle (mix of soil and husk) then a media of activated Alumina removes arsenic.
- Rainwater harvesting (RWH) -a) household level b) community level
- Long term measures pipe water supply.
- Immediate measures- deep tube Wells, tube Wells with attached ARP, HRF/SSF and domestic filter.
- Developed and tested by Bengal Engineering and Science University.
- Candle to remove iron (and arsenic) and activated Alumina cartridge to remove plastic.
- Rural sanitary Mart manufacturer for the production and promotion of domestic filter.
- Sale price 575 (up to per 50% subsidy for the BPL families).
- To get around 15,000 domestic filters sold providing arsenic safe water to estimated 90000 peoples.

Progress Made So Far

Several methods have been developed to remove arsenic from groundwater by the public health and engineering department government of West Bengal, All India Institute of hygiene and Public Health, tropical School of Medicine Calcutta and school of environmental studies Jadavpur University which are indicated as follows:

- Arsenic removal plant fitted directly with hand pump.
- Arsenic removal plant for piped water supply scheme.
- Surface water projects aiming to give safe drinking water are under construction.
- Domestic filters- various types domestic filters have been developed by the aih pH, PHED. This filters are fitted with sand as well as candle.
- School of environmental studies Jadavpur University have also developed such filters with specially manufactured candles, these candles are fitted with tablets of chemicals which are disposable.
- Amal arsenic removal units etc.

The above-mentioned scheme should be implemented adequately. But their application is less than necessary. People have very little knowledge about them, Public awareness is very necessary people in the study area are also being affected by the lack of awareness that they are drinking arsenic contaminated water.

Plate 4: Arsenic Free Water Tank and Tap

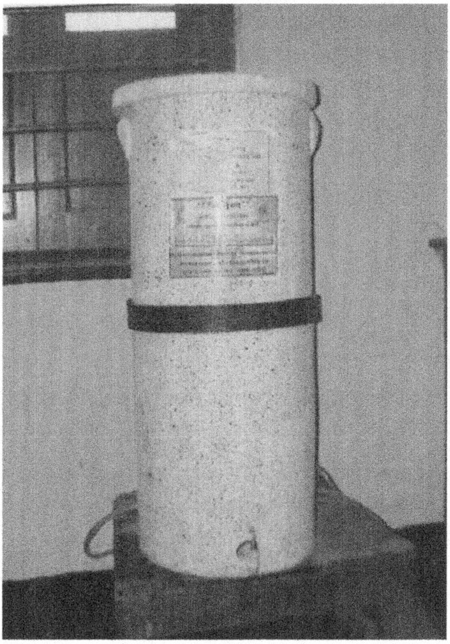

Plate 5: Domestic Water Filter
Source: Captured by Researcher

Conclusion

An urgent need is felt for arsenic free water for blocks and panchayats which is suitable and satisfactory to growing needs of people. More mitigation schemes are needed. Mass awareness is needed.

References

Amrita Bari (Issue April-June 2010), Magazine of CCDU.

Ahamad SA, Khan MH, Hoque M (2018), Arsenic Contamination in groundwater in Bangladesh: Implications and Challenges for healthcare policy, 1st Edition, 251-261.

http://cpheeo.gov.in/cms/manual-on-water-supply-and-treatment.php(Date- 10/06/2020)

Murshidabad Zila Parishad (2011): Water sample testing reports of Murshidabad 2011.

Sevavrata N.G.O (2011): water sample testing reports of Murshidabad 2011.

Shaban Abdul (2012) Applied Geography: A Research application for development, 1st Edition, Kolkata, 115-135.

Shankar Shiv, Shankar Uma, Sikha (2014), The Scientific World Journal, Arsenic contamination of groundwater: A review of sources, prevalence, Health, risks and strategies for mitigation,1st Edition.

Srivastava Anamika, Barla Anil, Yadav Himanshu, Bose Sutapa (2014), Frontiers in Environmental Science, Arsenic Contamination in Shallow groundwater and agricultural soil of Chakdaha, West Bengal, India,1st Edition.

References for Plates and Map

Plate 1: https://www.thenewleam.com/2017/09/arsenic-poisoning-water-rampant- cancer-outbreaks-north-india/

Plate 2: https://www.flickr.com/photos/simavi/5243262423

Plate 3: https://www.atsdr.cdc.gov/csem/csem.asp?csem=1&po=11

Map 1: https://www.semanticscholar.org/paper/Arsenic-contamination-of-groundwater-and-its-health-Rahman-Sengupta/843ececc3cdce101001dea4b0ee72144145cb0f9/figure/0

Part - III

Environmental Management and Practices

17

Application of Modern Geospatial Tools in Vegetation and Water Estimation: A Case Study of Burisuti Pathar Wetland Area of Assam

Dr. Krishna Das

Assistant Professor, Department of Geography, Pragjyotish College, Guwahati, Assam, India

Abstract

Geospatial technology is an emerging field now-a-days for studying and observing earth's natural resources, land use-land cover, spatial patterns of geographic phenomena, hazard and disaster events, changing earth's landscapes etc. Modern geospatial tools are integrated with advanced cartographic capabilities for addressing any geographic issue. These geospatial tools help researchers in estimating and mapping vegetation, waterbody, agricultural land etc. effectively using various geospatial techniques from recent satellite data. This study has also been carried out using modern geospatial tools and techniques to estimate vegetation and water from recent satellite imagery in a wetland area of Assam. This study is carried out using a Landsat 8 satellite imagery of January 2020 over the Burisuti pathar wetland area situated in Dhemaji district of Assam. The study has estimated vegetation and water cover areas in the study area through two classic remote sensing spectral indices. These indices include NDVI and NDWI. This study has applied open-source EOS

Land Viewer and QGIS tools for geospatial analyses. According to the results of this study, this has been found that there are more than fifty percentages of vegetation in the study area, while water cover area is about one percent proportion.

Keywords: NDVI, NDWI, Cluster Analysis, Modern Geospatial Tools, Burisuti Pathar.

Introduction

Geospatial tools integrate various technologies like remote sensing, geographic information system (GIS), global satellite navigation system (GNSS) and cartography for spatial decision-making using geographic data of varied nature through spatial analysis, modelling and geo-visualization. Now-a-days, modern geospatial tools are widely used by the researchers for natural resource assessment, environmental monitoring, agricultural development, disaster management, urban planning, traffic management, tourism development, etc. due to the multifaceted advantages of these tools (Trautmann & MaKinster, 2010; Zhichkina *et al*, 2020). Geospatial tools have the capabilities to manage the huge scale of geospatial data, and these tools can process and analyse geospatial data at cost-time effective manner. Modern geospatial tools also provide users with the one-stop solution right from satellite data download, pre-processing, analysis to geo-visualization functions in a single platform. Modern geospatial tools include online and cloud-based integrated geospatial platforms such as *Google Earth, ISRO's Bhuvan, Land Viewer Earth Observing System* with advance capabilities (Lawes *et al*, 2019). These modern geospatial tools are popular among researchers for various applications. Liang *et al* (2018) reviewed the levels of the different application of *Google Earth* in scientific publication, geosciences studies, environmental management studies and geographical studies. Muppalla *et al* (2019) addressed the web-mapping capabilities and functions of *ISRO's Bhuvan* platform. Kavats *et al* (2019) observed the crop harvesting monitoring based on satellite time series data using *Land Viewer Earth Observing System* platform. Similarly, this study has also been carried out with the research objective to show the application of modern geospatial tools in vegetation and water estimation over a wetland area of Assam (India).

The study is implemented in the *Burisuti pathar* wetland area of Assam as the case study site. Based on modern geospatial tools, the study assesses the vegetation and water in the study area. Vegetation and water are estimated from multi-spectral Landsat 8 OLI satellite imagery in the study area using geospatial tools (Feyisa *et al, 2014;* Gupta *et al,* 2018). Normalized Difference Vegetation Index (NDVI) is adopted in this study to estimate vegetation form the Landsat 8 data. The NDVI is the most common and widely used spectral remote sensing index for calculating vegetation greenness and status (Ashraf & Nawaz, 2015). On the other hand, Normalized Difference Water Index (NDWI) is used to estimate water from the Landsat 8 data. The NDWI is a classic and most frequently used spectral remote sensing index for estimating water from satellite imagery (Acharya *et al,* 2018). From these two indices of NDVI and NDWI, the vegetation and water cluster zones are derived for the study area respectively using the geospatial platform.

Methodology

The Study Area

The case study is confined to the Burisuti pathar wetland area of Assam. The Burisuti pathar wetland is situated in Dhemaji district of Assam to the north bank of the Brahmaputra river. The Burisuti pathar wetland area is located between 27.674° north to 27.721° north latitudes and 95.002° east to 95.091° east longitudes (Figure 1). The study area covers a geographical area of 27.00 km^2. The Burisuti pathar is an area surrounding the Burisuti riverine inland wetland.

Data Sources

The study is carried out using the U.S. Geological Survey (USGS) Landsat 8 Operational Land Imager (OLI) satellite imagery of 12th January 2020 for the winter season. This particular Landsat 8 imagery has path number 135 and row number 41 with 8 spectral bands of 30-meter resolution and 1 panchromatic band of 15-meter resolution (Du *et al,* 2014). Among these 9 spectral bands of Landsat 8 OLI data, only 3 bands are included in this study (Table 1).

Table 1: Landsat 8 OLI Imagery Bands Used in This Study.

Bands	Wavelength (micrometres)	Resolution (meters)	Coordinate reference system
Band 3 - Green	0.53-0.59	30	WGS 1984 datum UTM 46 North projection
Band 4 - Red	0.64-0.67	30	
Band 5 - Near Infrared (NIR)	0.85-0.88	30	

Source: https://www.usgs.gov/faqs/what-are-band-designations-landsat-satellites

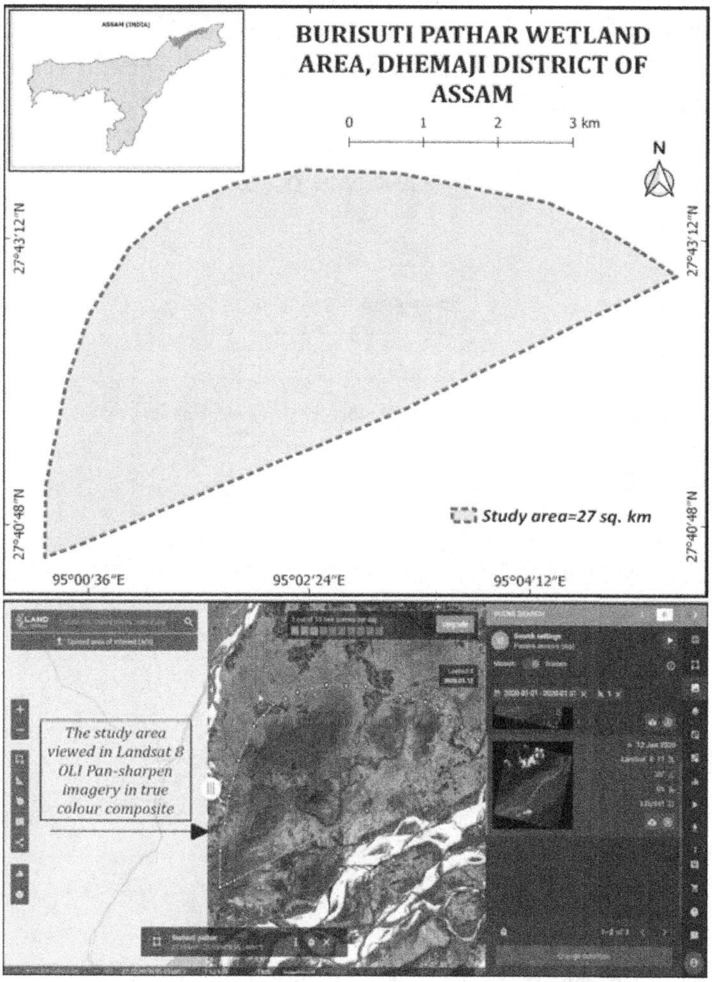

Figure 1: Location of The Case Study Area.

Methods

To estimate the vegetation and water using modern geospatial tools in the case study area, two spectral remote sensing indices are used. These two techniques of indices are NDVI and NDWI. These two indices are calculated from the Landsat 8 OLI satellite imagery of the study area.

NDVI is a band-ration technique which includes the near-infrared band and red band of multispectral satellite imagery to calculate the vegetation greenness. The calculated value range of NDVI is always between 0 to ± 1, where positive value represents healthy green vegetation and negative value represent non-vegetation cover (Wilson & Norman, 2018). The mathematical equation of NDVI is *NDVI=(NIR-RED)/(NIR+RED) or NDVI= (Band 5-Band 4)/ (Band 5+Band 4)* for Landsat 8 data used in this study.

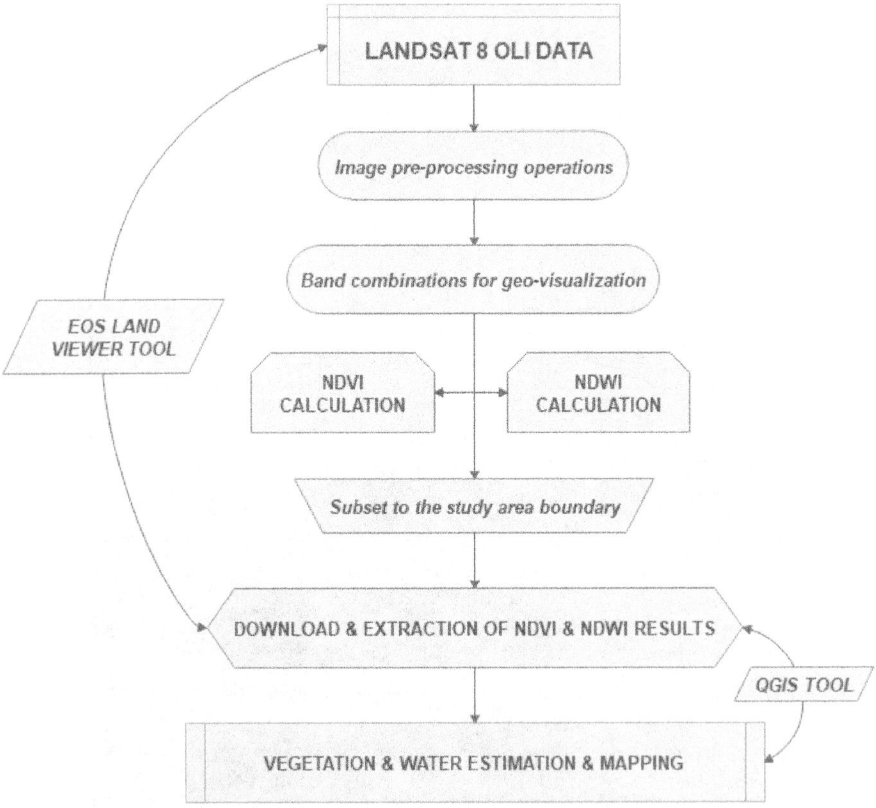

Figure 2: GIS Procedure Flowchart Adopted in This Study

NDWI is another band-ration technique which includes the green band and near-infrared band of multispectral satellite imagery to calculate the clear water bodies. The calculated value range of NDWI is always between 0 to ± 1, where positive value represents clear surface water bodies and negative value represent non-water cover areas (Liu *et al*, 2016). The mathematical equation of NDWI is *NDWI=(GREEN-NIR)/(GREEN+NIR) or NDWI= (Band 5-Band 4)/ (Band 5+Band 4)* for Landsat 8 data used in this study.

To perform cluster zoning of vegetation and water cover areas in the study area based on intensity values of NDVI and NDWI respectively, the NDVI and NDWI imageries are reclassified using minimum and maximum calculated values into three cluster zones (Das, 2017). These three cluster zones are classified following manual classification method in GIS.

The entire study has been performed using modern geospatial tools like *Google Earth, EOS Land Viewer Tool* and *Quantum GIS software package version 3.4* (Figure 2). The study area shapefile boundary is digitized and downloaded from the *Google Earth Pro* platform (https://earth.google.com/web). Right from Landsat 8 OLI imagery search for the study area to image pre-processing, true colour visualization, band combination, NDVI and NDWI generation and calculation to final NDVI and NDWI imageries with statistics extortion and download are performed using the *EOS Land Viewer Tool* (Kolodiy & Pidlypna, 2020). The *EOS Land Viewer Tool* is an online geospatial platform the has various image-processing, analysis and visualization capabilities on any part of the world for free (https://eos.com/landviewer/). The users only need to register for this one-stop geospatial operations. NDVI and NDWI cluster analyses, maps output preparation and map-statistics calculations are performed using open-source *QGIS* desktop software platform (https://qgis.org/en/site/).

Results and Discussion

To identify the vegetation from the Landsat 8 OLI imagery of 2020, NDVI is generated using *EOS Land Viewer Tool* for the Burisuti pathar area. The *EOS Land Viewer* platform can automatically generate NDVI visualization for the selected satellite imagery. This generated NDVI map of the study area from the *EOS Land Viewer* platform is downloaded and visualized in GIS using grayscale colour ramp showing the calculated pixel values of NDVI. In this NDVI map of the study area, positive NDVI value pixels are shown in white colour and negative NDVI value pixels are shown in grey to black colour (Figure 3). These positive NDVI pixels in white colour represent the thick, dense and green vegetation cover of the study area. The calculated NDVI map values range between *-0.319* to *0.809* with the mean NDVI value of *0.506*. Similarly, to identify the waterbodies from the Landsat 8 OLI imagery of 2020, NDWI is generated using *EOS Land Viewer Tool* for the Burisuti pathar area. The *EOS Land Viewer* platform is also capable offer automatic generation of NDWI visualization for the selected satellite imagery. This generated NDWI map of the study area from the *EOS Land Viewer* platform is downloaded and visualized in GIS using grayscale colour ramp showing the calculated pixel values of NDWI. In this NDWI map of the study area, positive NDWI value pixels are shown in white colour and negative NDWI value pixels are shown in grey to black colour (Figure 4). These positive NDWI pixels in white colour represent the clear water bodies of the study area. The calculated NDWI map values range between *-0.513* to *0.690* with the mean NDWI value of *-0.450*.

These NDVI and NDWI maps of the study area have provided only the geo-visualization of the calculated pixels. Form these maps, an overview of the distribution of the positive and negative pixel can be understood. Therefore, cluster analysis has been performed in these NDVI and NDWI maps of the study area in delineating and estimating the actual vegetation and water areas. The cluster analysis is performed to classify the study area into distinct intensity zones based on NDVI and NDWI range values. Both NDVI and NDWI maps are individually classified into intensity cluster zones using manual data grouping technique in GIS. The NDVI map of the study area has been classified into three distinct intensity cluster zones based on the minimum and maximum NDVI value range. These three intensity cluster zones are *High, Moderate* and *Low*. The high-intensity NDVI zone has the pixels with more than *0.50* NDVI value and this zone represents the dense, thick and healthy vegetation in the study area. The low-intensity NDVI zone represents non-vegetation areas having negative NDVI pixels. While the moderate-intensity NDVI zone represents both sparse vegetation and non-vegetation areas (Figure 5).

EOS LAND VIEWER NDVI VISUALIZATION OF THE STUDY AREA

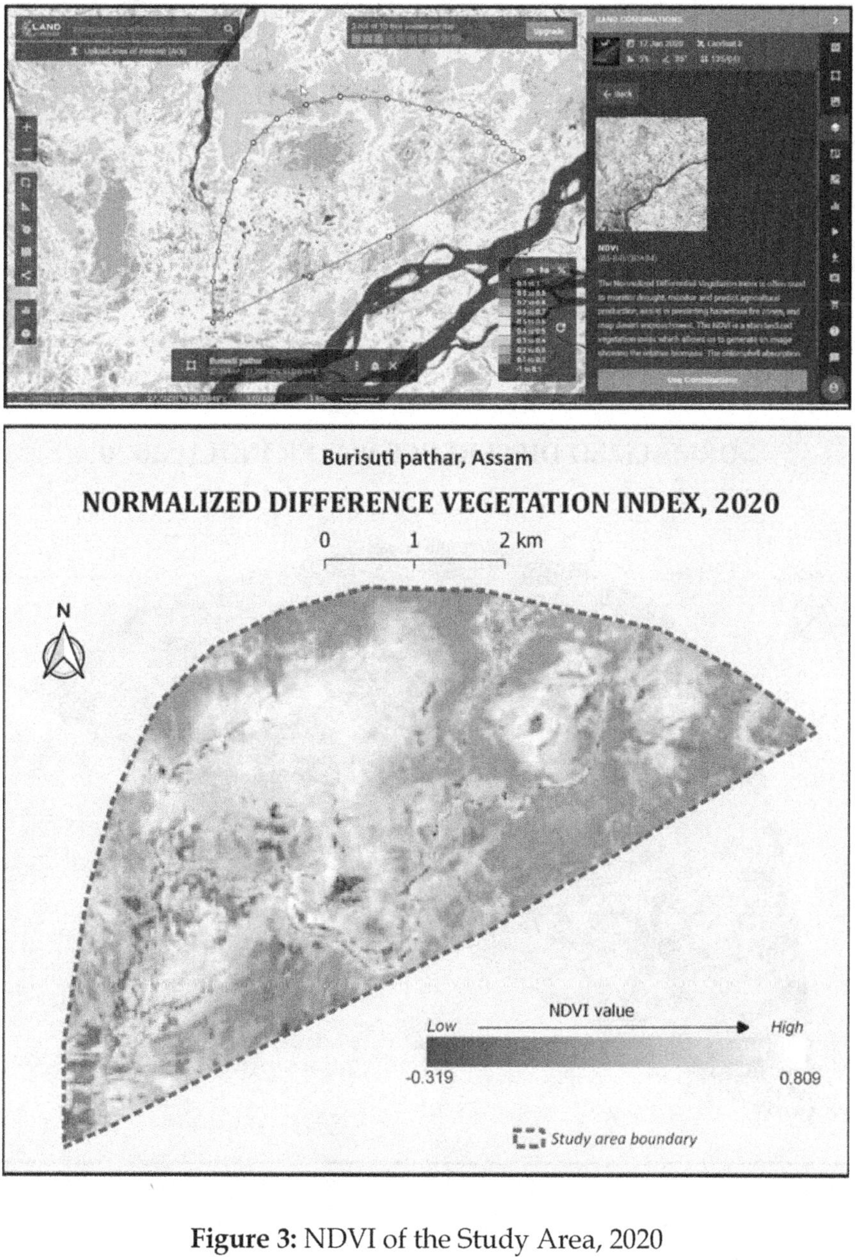

Figure 3: NDVI of the Study Area, 2020

EOS LAND VIEWER NDWI VISUALIZATION OF THE STUDY AREA

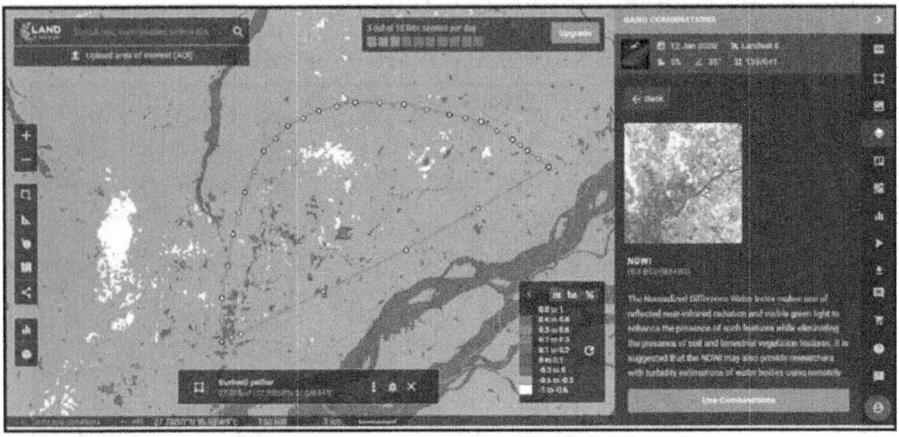

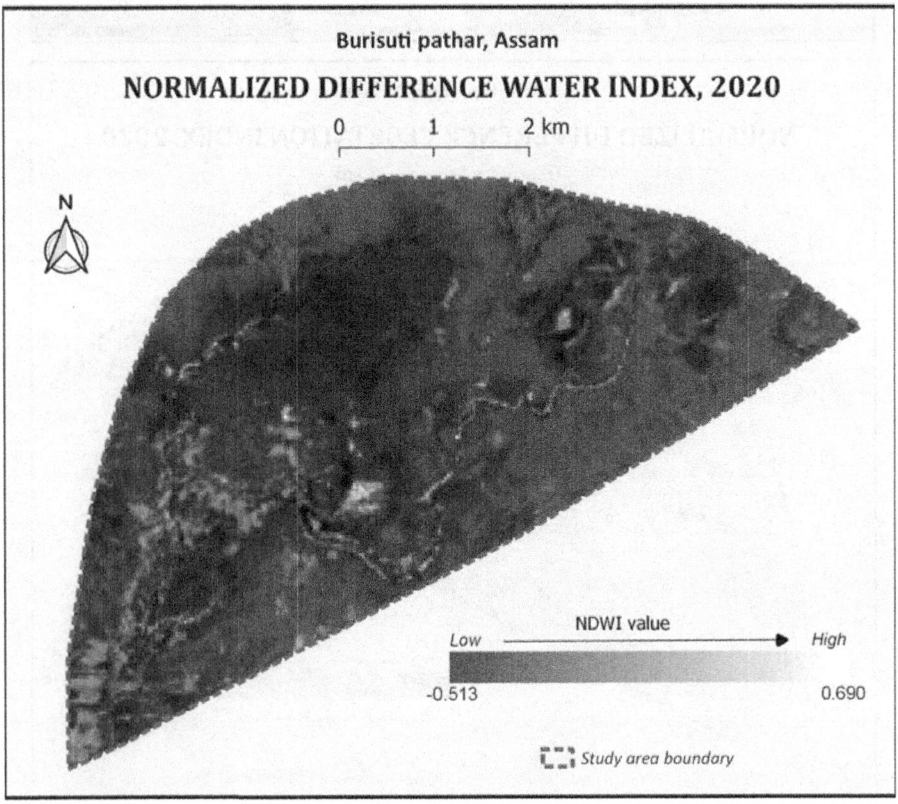

Figure 4: NDWI of the Study Area, 2020

Table 2: NDVI Intensity Cluster Statistics, 2020

Cluster zone	NDVI Intensity range	Total pixels	Total area (km^2)	Proportion (%)
Low	-0.31 to 0.20	416	0.3744	1.39
Moderate	0.20 to 0.50	13540	12.186	45.21
High	0.50 to 0.80	15994	14.3946	53.40
Study area total		29950	26.955	100.00

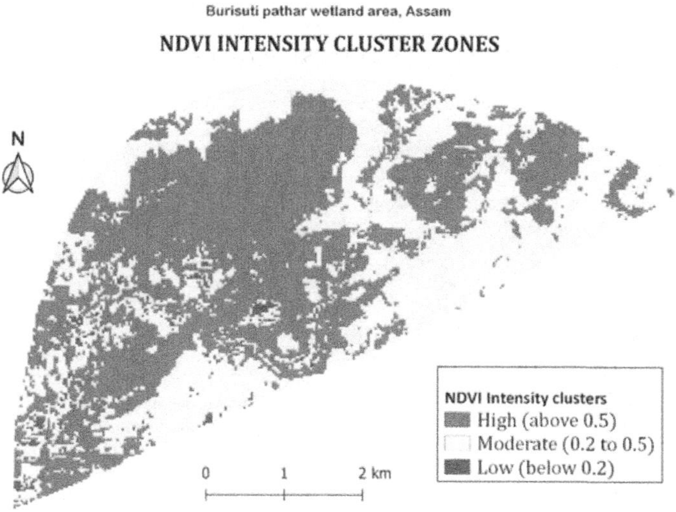

Figure 5: NDVI Intensity Cluster Zones, 2020

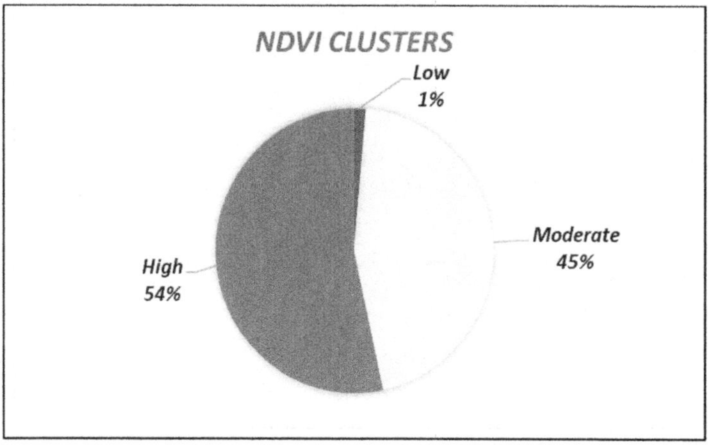

Figure 6: NDVI Intensity Cluster Proportions in The Study Area

Table 3: NDWI Intensity Cluster Statistics, 2020

Cluster zone	NDWI Intensity range	Total pixels	Total area (km^2)	Proportion (%)
Low	-0.69 to -0.50	10149	9.1341	33.89
Moderate	-0.50 to -0.10	19480	17.532	65.04
High	-0.10 to 0.51	321	0.2889	1.07
Study area total		29950	26.955	100.00

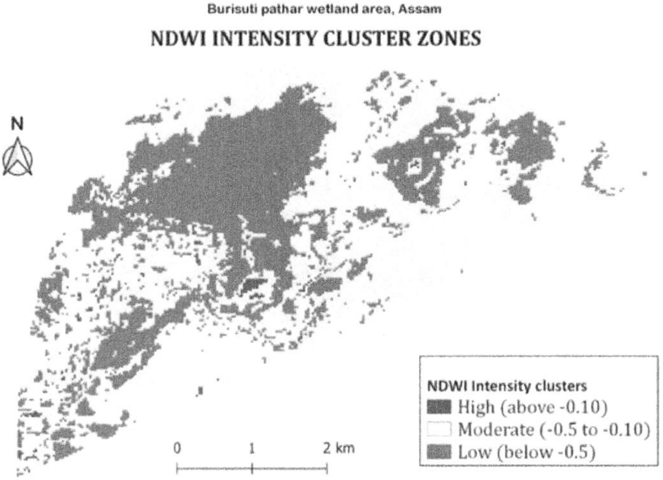

Figure 7: NDWI Intensity Cluster Zones, 2020

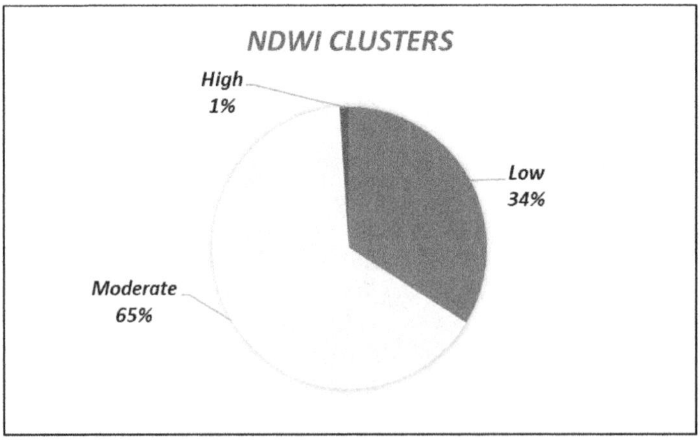

Figure 8: NDWI Intensity Cluster Proportions in The Study Area

The statistics generated from this NDVI intensity zones of the study area have revealed that there are *14.39* km^2 of the high-intensity zone, *12.18* km^2 of the moderate-intensity zone and less than *1* km^2 of the low-intensity zone in the study area (Table 2). This has been also observed that there is a maximum proportion of high NDVI intensity zone in the study area which shows that the study area has about *53* percent of vegetation cover (Figure 6). Now, to estimate the water in the study area, the NDWI map has been classified into three distinct intensity cluster zones based on the minimum and maximum NDWI values. These three intensity cluster zones are *High, Moderate* and *Low*. The high-intensity NDWI zone has the pixels with more than *-0.10* NDWI value and this zone represents the clear and open water in the study area. The low-intensity NDWI zone represents non-water areas having less than *-0.50* NDWI pixels. While the moderate-intensity NDWI zone represents both swamp and non-water areas (Figure 7). The statistics generated from this NDWI intensity zones of the study area have revealed that there is about *1* km^2 of the high-intensity zone, *17.53* km^2 of the moderate-intensity zone and less than *9.13* km^2 of the low-intensity zone in the study area (Table 3). This has been also observed that there is minimum proportion of high NDWI intensity zone in the study area which shows that the study area has about *1* percent of clear and open water cover area (Figure 8).

Conclusion

This study has adopted NDVI technique for estimation of vegetation cover and NDWI technique for estimation of clear water cover in the study area. The study has observed that there is more than fifty percent proportion of vegetation and more than one percent proportion of water in Burisuti pathar wetland area of Assam. Based on NDVI and NDWI cluster analyses, the study has revealed that the north-central part of the study area is mostly covered with thick vegetation, while water body areas are distributed in small and compact patches in the western part of the study area. The study has also observed effective application of the modern geospatial tool *EOS Land Viewer* in extracting the NDVI and NDWI observations based on January 2020 Landsat 8 OLI imagery of the study area. Finally, it can be concluded that the entire study has provided relevant findings on vegetation and water estimation in the study area and this study will also help other researchers in carrying out similar studies using modern geospatial tools for natural resource and environmental management observations of wetlands.

Acknowledgement

The author acknowledges the Earth Observing System Group of Land Viewer and the Quantum GIS Project developer group for providing these modern open-source geospatial tools for analyses and observations.

References

Acharya, T. D., Subedi, A., & Lee, D. H. (2018). Evaluation of water indices for surface water extraction in a Landsat 8 scene of Nepal. *Sensors, 18*(8), 2580.

Ashraf, M., & Nawaz, R. (2015). A comparison of change detection analyses using different band algebras for baraila wetland with Nasa's multi-temporal Landsat dataset. *Journal of Geographic Information System, 7*(01), 1.

Das, K. (2017). NDVI and NDWI based Change Detection Analysis of Bordoibam Beelmukh Wetlandscape, Assam using IRS LISS III data. *ADBU Journal of Engineering Technology, 6*(2).

Du, Z., Li, W., Zhou, D., Tian, L., Ling, F., Wang, H., & Sun, B. (2014). Analysis of Landsat- 8 OLI imagery for land surface water mapping. *Remote sensing letters, 5*(7), 672-681.

Feyisa, G. L., Meilby, H., Fensholt, R., & Proud, S. R. (2014). Automated Water Extraction Index: A new technique for surface water mapping using Landsat imagery. *Remote Sensing of Environment, 140*, 23-35.

Gupta, K., Mukhopadhyay, A., Giri, S., Chanda, A., Majumdar, S. D., Samanta, S., & Hazra, S. (2018). An index for discrimination of mangroves from non-mangroves using LANDSAT 8 OLI imagery. *MethodsX, 5*, 1129-1139.

Kavats, O., Khramov, D., Sergieieva, K., & Vasyliev, V. (2019). Monitoring Harvesting by Time Series of Sentinel-1 SAR Data. *Remote Sensing, 11*(21), 2496.

Kolodiy, P., & Pidlypna, M. (2020). The Improvement of the Agricultural Yields Forecasting Model Using the Software Product "Land Viewer". *Geomatics and Environmental Engineering, 14.*

Lawes, R. A., Chai, J., Chen, Y., Donohue, R. J., Mata, G., Hochman, Z., & Butler, R. (2019). Monitoring Grain Production across the Australian Continent. In *Proceedings of the Agronomy Australia Conference, Wagga Wagga, Australia.*

Liang, J., Gong, J., & Li, W. (2018). Applications and impacts of Google Earth: A decadal review (2006–2016). *ISPRS Journal of Photogrammetry and Remote Sensing, 146,* 91-107.

Liu, Z., Yao, Z., & Wang, R. (2016). Assessing methods of identifying open water bodies using Landsat 8 OLI imagery. *Environmental Earth Sciences, 75*(10), 873.

Muppalla, A. R., Naresh, N., Kumar, P. A., Bothale, V. M., Joshi, A. K., & Prasad, A. V. V. (2019, October). Study of open source web map rendering software for rendering Bhuvan High Resolution Satellite data. In *2019 IEEE Recent Advances in Geoscience and Remote Sensing: Technologies, Standards and Applications (TENGARSS)* (pp. 141-145). IEEE.

Trautmann, N. M., & MaKinster, J. G. (2010). Flexibly adaptive professional development in support of teaching science with geospatial technology. *Journal of Science Teacher Education, 21*(3), 351-370.

Wilson, N. R., & Norman, L. M. (2018). Analysis of vegetation recovery surrounding a restored wetland using the normalized difference infrared index (NDII) and normalized difference vegetation index (NDVI). *International Journal of Remote Sensing, 39*(10), 3243-3274.

Zhichkina, L. N., Nosov, V. V., Zhichkin, K. A., Aydinov, H. T., Zhenzhebir, V. N., & Kudryavtsev, V. V. (2020). Satellite monitoring systems in forestry. In *Journal of Physics: Conference Series* (Vol. 1515, p. 032043).

18

Cross-Disciplinary Approach Between Chemistry and Geography for Environmental Problems: An Overview

Mominul Sinan

Assistant Professor, Department of Chemistry,
Charuchandra College, Kolkata, West Bengal, India

Abstract

Environmental issues in the 21st Century have changed dramatically in the last years. Developments has caused Environmental resources are under the pressure. Environmental issues related to natural resources, biodiversity, sustainability, and human health are on the top on the list of priorities and concern of national and international environmental protection organizations. Scientists around the world from different fields are trying to address these environmental issues through sustainable ways so that the needs of people can be keep maintained without destabilising ecological integrity. Environmental science is an interdisciplinary academic field that integrates physical, biological and information sciences to the study of the environment, and the solution of environmental problems. Environmental geography researches the movement, transformation, and effect of pollutants under different spatial and temporal scales in the geographical environment, whereas chemical sciences help to analyze adverse effect of pollutant in environment and tries to address some solutions regarding the problems. Some environmental issues have been discussed here on chemical and geographical perspective.

Keywords: Chemistry, Environmental Geography, Sustainable Development.

Introduction

Globalization has come up with new economic, social and technological changes on a global scale. Developments has caused Environmental resources are under the pressure. Any undesirable and unwanted change in physical, chemical and biological characteristics of air, water and soil which are harmful for living organisms can be termed as Environmental pollution (Wong, 2012). The sustainability (Polasky, 2019) can only exist if all environmental, economic and social issues are addressed together. Sustainability recognises the integration of humans and nature as an inescapable reality thus a new form of political, environmental and social mobilizations appear in developed and developing countries. Scientists around the world from different fields are trying to address these environmental issues through sustainable ways so that the needs of people can be keep maintained without destabilising ecological integrity. Emerging environmental problems are the new challenges to international and national environmental protection institutions.

Environmental Organisations (Pariona, 2017)

Environmental issues related to natural resources, biodiversity, sustainability, and human health are on the top on the list of priorities and concern of national and international environmental protection organizations. These organizations played important role in information and understanding of the threats, promoted investigations and follow up studies from environmental trends through their research. Some of these international organisations are United Nations Environment Programme (UNEP), Global Green Growth Institute (GGGI), and Inter-governmental Panel on Climate Change (IPCC). International Union for Conservation of Nature (IUCN), Environmental Protection Agency (EPA), European Environment Agency (EEA), European Environmental Bureau (EEB), Partnerships in Environmental Management for the Seas of East Asia (PEMSEA), etc. Non-governmental organizations (NGOs) also playing an active role in this program.

The UN Environment's Sixth Global Environmental Outlook (GEO-6) in 2019 (UN Environment, 2019) under the theme "Healthy Planet, Healthy People", offers a rigorous analysis of Earth's prospects for a healthy future. Important environmental issues that the Earth will be facing in the next decades have been discussed here. Governments and other stakeholders take decision based on the information given by GEO. Some action are reducing land degradation, biodiversity loss, and air, land and water pollution; improving water management and resource management; climate change mitigation and adaptation; resource efficiency; addressing decarbonisation, decoupling and detoxification; and the prevention and management of risk and disasters. Those all require more ambitious and effective policies, including sustainable consumption and production, greater resource efficiency and improved resource management, integrated ecosystem management, and integrated waste management and prevention.

Interdisciplinary Environmental Science

Environmental science is an interdisciplinary (Mendeley, 2018) academic field that integrates physical, biological and information sciences (including ecology, biology, physics, chemistry, plant science, zoology, mineralogy, oceanography, limnology, soil science, geology and physical geography, and atmospheric science) to the study of the environment, and the solution of environmental problems. Here some discussion has been done on chemistry and geographic perspective.

Improving waste management (David C. Wilson, 2007), including recycling and end of-life management is a major challenge for environmental chemists as well as environmental geographers. As the hazardous waste mainly contains chemicals, heavy metals, carcinogens, pathogens, radioactive or other toxic materials which are the by-products of any process. There are many analytical techniques (Goulden, 1978) for analysing the level of pollution in chemistry. Environmental Chemists are working to reduce, reuse or recycle (Dijkers, 2019) the waste as much as possible. A subcategory in chemistry, popularly known as green chemistry (Sinan, 2018), is wholly dedicated for this purpose.

On the other hand a sub-discipline of geographical sciences, environmental geography (Concordia, 2020) researches the movement, transformation, and effect of pollutants under different spatial and temporal scales in the geographical environment by taking environmental problems as the orientation and geographical method. Some methods in environmental geography are Ecological Modelling, Remote Sensing, Multivariate Statistics, Time-series Analysis, Geochronology, Isotope Biogeochemistry etc. They use physical process as well as demographic, cultural, and economic processes to address environmental problems. It also integrates methods from other disciplines. Interdisciplinary development in environmental geography help researcher to research the environmental processes and behaviours of pollutants to show the multi-scale, multi-interface, and multi-process characteristics. Both fieldwork and laboratory work are the key part of studying environmental science. Testing, analysing and interpretation are the basic tools for environmental analysis. Some issues related to chemistry and geography is discussed below.

Heavy Metal

Heavy metal (Singh, 2011) always has been the major pollutant researched by environmental geography. Air pollutions from heavy metals, along with their adsorption and desorption process into soil; their toxicity, bio-availability and risk assessment in soil, water and sediment and the remediation of contaminated soil have reflected by the recent research. We know that metals are essential for a wide range of industries like mechanical engineering, transport, aerospace, construction, packaging, electricity and energy, consumer electronics, medical devices, etc. The European Commission indentified the most critical metals: Heavy rare earths, Light rare earths, Magnesium (Mg), Antomony (Sb), Tungstem (W), Gallium (Ga), natural graphite (C), Germanium (Ge), Indium (In), Silicon (Si), Lithium (Li), Fluorspar (CaF_2, calcium fluoride), Cocking coal (C), Phosphate rock (P_2O_5), Niobium (Nb), Beryllium (Be), Platinum (Pt), Chromium (Cr), Cobalt (Co), Borates (large number of boron-oxygen compounds usually containing oxyanions). Although the afore mentioned heavy metals can be found (as hydroxides, oxides, sulphides, sulphates, phosphates, silicates) in traces, they still cause serious health problems to human and other mammals (Herawati, 2000).

Heavy metals presence in the environment leads to a number of adverse in all spheres of the environment i.e. hydrosphere, lithosphere, biosphere and atmosphere. Until the impacts are dealt with, health and mortality problems break out, as well as the disturbance of food chains. Among the heavy metals, Zn, Ni, Co, and Cu are relatively more toxic to plants, and As, Cd, Pb, Cr and Hg are relatively more toxic to higher animals (McBride, 1994).

During more than a century the use and extraction of mineral resources increased (Resilience, 2017) exponentially (Figure 1a) with an average between 3-4% annually. As there are limited amounts of mineral resources in the Earth's crust a depletion of some very important natural resources is surprising.

Organic pollutants are also major concern as they have some unique properties likes high lipid solubility, stability, lipophilicity and hydrophobicity, thus they are accumulated (Van Ael, 2012) in the different spheres of the environment, and causing toxicological effects. Few of them are polychlorinated biphenyls (PCBs), polybrominated diphenyl ethers (PBDEs), polycyclic aromatic hydrocarbons (PAHs), pesticides, petroleum and organochlorine pesticides (OCPs) (El-Shahawi, 2010). More pollutants with close relationships to human health have become research topics.

Hazardous/Toxic Waste

Hazardous or toxic wastes (Wilson, 2007) in any form can cause substantial threats to environmental resources like air, water, soil to human health. This waste may come from verities of process like municipal and households generated waste, waste from manufacturing, farming, water treatment systems, construction, automotive garages, laboratories, hospitals, and other industries. Disposal of hazardous or toxic substance (WOLTERS, 2019) is a major environment issue. Safely disposal of hazardous waste is expensive so some people decide to just ship it to other poorer parts of the world. Toxic waste can harm ecosystem through soil, air and water pollution. Some of this disposal pollution discussed below.

Soil Pollution

Polluted water streams, low-density polyethylene (LDPE) plastic bags and industrial sources like fly ash and chemical residues as well as pesticides, herbicides, fertilizers etc are the common sources of soil pollution. Different types of pollutant (Swartjes, 2011) from chemistry point of views shown in Figure 1b. Electronic waste (discard telephones, computers, etc) is also causing threat to environment.

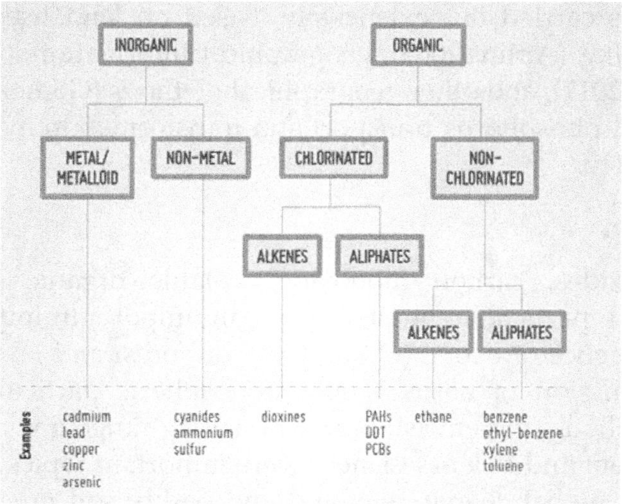

(a)

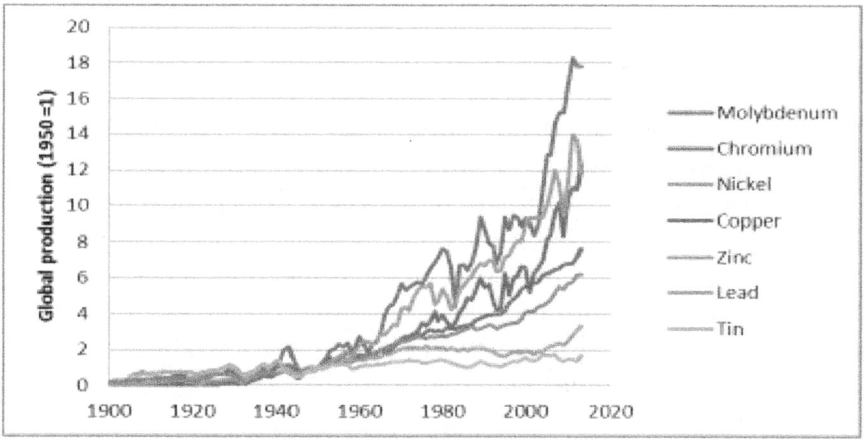

(b)

Figure 1: (a) Different Soil Pollutant; (b) Trends in the Annual Extraction of Seven Strategic Mineral Commodities that are Used in Many Industrial and Commercial Industries

Water Pollution

Eutrophication (Yang, 2008) and biological effects, speciation, desorption, transport and transformation, toxicity and bioavailability of pollutants in sediments formed the main content of research in geography. Watershed-scale non-point source pollution load estimation research was carried out extensively, based on land use type using remote sensing (Aplin, 2005), geographical information system (GIS) (Srivastava, 2011), and other geographical methods (Gomez, 2010). The nitrogen and phosphorus transport and transformation in water body studied widely.

Air Pollution

Nitrogen oxides, carbon monoxide, volatile organic compounds (VOCs), and particulate matter have an important impact on air quality. Influence of ozone (O_3) and SO_2 on forest and lake ecological systems, emission of aerosols and atmospheric particulates (PM_{10}, $PM_{2.5}$), greenhouse gas emission, acid rain, long-range transport, source apportionment and risk assessment were important topics of research. Research on global change, wetland and paddy soil greenhouse gas emission research appeared recently, becoming a popular topic in environmental geography research.

Chemical Techniques for Pollution Assessment

Laboratory analytical instrumentation has advanced (Hussain, 2020) considerably in recent years, and is now commercially available. Different Instruments and methods (Kumar, 2017) such as GC, GC/MS, HPLC, IC, AA, XRF, INNA, plasma induced spectroscopy, proton induced X-ray emissions, UV-visible and IR spectroscopy, constant energy synchronous luminescence, chemical luminescence, electrochemical, and light scattering are available. These techniques have been applied to the analyses of bulk and trace constituents of contaminants in the biosphere. Spectrophotometry is a sensitive analysis technique for measurement of very low analyte concentration. Atomic absorption spectrophotometry (AAS) is a very versatile and widely used technique as it helps in determining a large number of metals in soil, water, air, and blood, present at fairly low concentration (parts per billion or parts per million). Flame photometry, also known as atomic emission spectrometry, is an environmental analysis technique used to determine the concentration of alkali metal salts, mainly Sodium (Na), Calcium (Ca), Potassium (K), Lithium (Li) and Caesium (Cs) with the help of a photoelectric flame photometer. Inductively Coupled Plasma Mass Spectroscopy (ICP-MS) ICP-MS is a combination of inductively coupled plasma which is considered as an ideal source for the generation of ions, and mass spectrometry is for the detection of ions, a technique based on the measurement of the mass of an element. IR spectroscopic technique used for the functional group determination a compound. X-ray diffraction or XRD is a non destructive analytical technique pioneered for the identification of crystalline materials.

Analysis of Driving Factors for Disciplinary Development

Cross-disciplinary approaches (Game, 2018) to Global Environmental Health are essential to address environmental health threats within and beyond national boundaries, taking into account the links among health, environment and socio-economic development. Environmental geography, sub-discipline of geographical sciences, researches the movement, transformation, and effect of pollutants under different spatial and temporal scales in the geographical environment by taking environmental problems as the orientation and geographical method as the chief means. It also integrates methods from other disciplines.

Environmental geography always considers heavy metal as major pollutant research. Emission and air pollution of heavy metals, adsorption and desorption of heavy metals in soil; toxicity, bio-availability, and risk assessment of heavy metals in air, soil, water and sediment, and the remediation of contaminated soil have been a major research topic. In addition to heavy metals, the pollutants of most concern have changed from PAHs, PCBs and dioxins to emerging pollutants such as PFCs, EDCs, PPCPs, antibiotics, and nanoparticles. Again Environmental biotechnology in particular is the application of processes for the protection and restoration of the quality of the environment. Environmental biotechnology can be used to sense, prevent and remove the emission of pollutants into the environment in a number of ways. There are various major different types of applications of environmental biotechnology. Environmental biotechnology is one of the fastest growing and most practically useful scientific fields. Research into the genetics, biochemistry, and physiology of exploitable microorganisms is rapidly being translated into commercially available technologies for reversing and preventing further deterioration of the earth's environment.

Conclusion

Environmental issues and emerging global challenges in the 21st Century have changed dramatically in the last years. Humankind has witnessed urgent pressures from the climatic changes and environmental pollution challenges by emerging pollutants. Despite the increasing numbers of international treaties, national and international regulations to reduce air pollution and water contamination by toxic and hazardous chemicals, a high proportion of pollutants are still discharged in air, sediment, rivers, lakes and open seas, oceans. Without geography's contributions, modern environmental science would not exist (William, 2004). Environmental geography is to address an increasing number of environmental issues. Contributions made by geography to the environmental sciences include integrated approaches that account for space, dimension, and distribution, and the concept that landscapes arise from interactions between man and nature. Extensive research into increasingly complex environmental issues in environmental geography has led to a cross-disciplinary approach using research methods from fields including environmental chemistry, atmospheric chemistry, geochemistry, biology, and environmental engineering.

Acknowledgement

I am thankful to my colleagues for their constructive suggestions during the manuscript preparation.

References

Aplin, P. (2005). Remote sensing: ecology, *Progress in Physical Geography,* 29(1), 104–113.

Brodin, E. M. & Avery, H. (2020) Cross-Disciplinary Collaboration and Scholarly Independence in Multidisciplinary Learning Environments at Doctoral Level and Beyond.

Concordia, (2020) Why study Environmental geography, Retrieved on February 29, 2020 from https://www.concordia.ca/artsci/geography-planning-environment/programs/undergraduate/environmental-geography.html#:~:text=Environmental%20geography%20is%20the%20branch,humans%20and%20the%20natural% 20world

Dijkers, M. (2019). Reduce, reuse, recycle: good stewardship of research data, *Spinal Cord,* 57, 165–166.

El-Shahawi, M.S., Hamza, A., Bashammakhb, A.S., Al-Saggaf, W.T. (2010). An overview on the accumulation, distribution, transformations, toxicity and analytical methods for the monitoring of persistent organic pollutants. *Talanta*. 80, 1587-1597.

Game, E. T., Tallis, H., Olander, L., Alexander, S. M., Busch, J., Cartwright, N., Kalies, E. L., Masuda, Y. J., Mupepele, A-C., Qiu, J., Rooney, A., Sills, E. & Sutherland, W. J. (2018).

Gomez, B. & Jones III, J. P. (2010). *Research Methods in Geography A Critical Introduction,* Blackwell Publishing Ltd, United Kingdom.

Goulden, P. D. (1978). Environmental pollution analysis. United States: N. p., Web.

Herawati, N., Suzuki, S., Hayashi, K., Rivai, I.F. & Koyoma, H. (2000). Cadmium, copper and zinc levels in rice and soil of Japan, Indonesia and China by soil type. *Bulletin of Environmental Contamination and Toxicology*. 64, 33-39.

Hussain, C. M. & Keçili, R. (2020). *Modern Environmental Analysis Techniques for Pollutants,* Elsevier.

Katsouyann, K. (2013). *Geographical distribution of air*, in K., Cohen et al. (eds.), *Air Pollution and Cancer, IARC scientific,* Publication no. 161, France, 13-27.

Kumar, R., Pathak, A., Chauhan, M. & Negi, S. (2017). *Measurement of Environmental Pollution: Types and Techniques*, in R. Kumar et al. (eds.), *Advances in Environmental Biotechnology*, Springer, 13-37.

Mendeley, (2018). Studying Environmental Science: What is it like and where can it take you? Retrieved on February 29, 2020 from https://www.mendeley.com/careers /article/environmental-science/

McBride, M.B. (1994). Environmental chemistry of soils. New York, Oxford University Press.

Minerva (2020). Cross-discipline evidence principles for sustainability policy, Nature Sustainability, 1, 452–454. https://doi.org/10.1007/s11024-020-09397-3

Pariona, A. (2017). Major International Organizations Fighting Environmental Destruction Retrieved on February 29, 2020 https://www.worldatlas.com/articles/major-international-environmental-organizations-operating-in-the-world-today.html

Polasky, S., Klingc, C.L., Levina, S.A., Carpentera, S.R., Daily, G.C., Ehrlich, P.R., Heal, G.M. & Lubchencoa, J. (2019) Role of economics in analyzing the environment and sustainable development, PNAS, 116(12), 5233–5238.

Resilience, (2017). *An Update on Mineral Depletion: Do we Need Mining Quotas?* Retrieved on February 29, 2020 from https://www.resilience.org/stories/2017-01-17/an-update-on-mineral-depletion-do-we-need-mining-quotas/

Sinan, M. (2018). Sustainable Development and Green Chemistry: A *Review. International Journal of Research on Social and Natural Sciences*, 3 (2), 48-67.

Singh, R., Gautam, N., Mishra, A. & Gupta, R. (2011) Heavy metals and living systems: An overview, *Indian Journal of Pharmacology*, 43(3), 246-253.

Srivastava, P.K., Mukherjee, S., Gupta, M. *et al.* (2011). Characterizing Monsoonal Variation on Water Quality Index of River Mahi in India using Geographical Information System. *Water Qual Expo Health*, 2, 193–203.

Swartjes, F.A., ed. (2011). *Dealing with Contaminated Sites*. Dordrecht, Springer, Netherlands.

UN Environment, (2019). Global Environment Outlook 6, Retrieved on February 29, 2020 from https://www.unenvironment.org/resources/global-environment-outlook- 6

Van Ael, E., Covaci, A., Blust, R. & Bervoets, L. (2012). Persistent organic pollutants in the Scheldt estuary: Environmental distribution and bioaccumulation. *Environmental International*. 48, 17-27.

William, M. M. & John, G. J. (2004). Environmental Geography: Science, Land Use, and Earth Systems, 3rd Edition, John Wiley & Sons, New York.

Wilson, D. C. (2007). Development drivers for waste management, *Waste Manage Res*, 2007, 25, 198–207.

WOLTERS, C. (2019). *Toxic waste, explained*, Retrieved on February 29, 2020 from https://www.nationalgeographic.com/environment/global-warming/toxic- waste/

Wilson, D. C. (2007). Development drivers for waste management, *Waste Manage Res*, 25, 198–207.

Wong, M.H. (2012) *Environmental Contamination: Health Risks and Ecological Restoration.*

United States of America: Taylor & Francis Group.

Yang, X., Wu, X., Hao, H. et al. (2008). Mechanisms and assessment of water eutrophication. *J. Zhejiang Univ. Sci. B*, 9, 197–209.

19

Interweaving Environment, Ecology and Feminism: Ecofeminism as The Means of Environment Conservation

Swati Sharma
Former Research Scholar, M.Phil., Department of History,
Babasaheb Bhimrao Ambedkar University, Lucknow,
Uttar Pradesh, India

Abstract

Growing population, technological advancement and over-exploitation of natural resources are putting great strain on environment and disturbing the ecological balance. Women are considered more closer to the nature because they play a critical role in managing natural resources for both family and community and are most affected by environmental degradation. Therefore, conservation and protection of natural resources cannot be done without involving women in planning and training for values regarding conservation of environment. Role of different women's movements and their struggle to preserve our environment will be analyzed in this work. With changes in environmental and ecological activism, the ways in which feminist activism have influenced the both have shaped a new ideology which is known as ecofeminism. In such a situation, ecofeminism has become the driving force for ecological movements around the world and is realised in terms of a theoretical understanding of ecological ethics as well as feminism. In its essential rendering it links women to the environment as they share common biological destinies that are being exploited by a culture that is dominated by patriarchy and men. This paper makes an attempt to undertake evaluation of ecofeminism in assessing the ways by which it is framed and understood in the emerging thrust towards ecological activism. This work analyses various grass-roots level movements and talks about eminent women environmentalists (such as Amrita Devi, Gaura Devi, Vandana Shiva, Maneka Gandhi and Radha Bhatt) and their contribution regarding this blended ideology of activism with everyday experiential realities.

Keywords: Environment, Environment Movement, Activism, Women, Ecofeminism, Natural Resources.

Introduction

The word 'environment' is derived from the French word 'environmer' which means surround. Environment is best described as the sum total of all living and non-living components (such as plants, animals, water, soil, air, etc.) which surrounds, influences and affects an organism during its life time. The study of relationship of living organisms with each other and their environment is known as ecology. In today's scenario, the two most important threats to the environment are- overpopulation and the destruction or exhaustion of natural resources.

As per the excerpt quoted by Francis d'Eaubonne these two threats were the outcome of two parallel discoveries which gave men power fifty centuries ago; the ability to plant seed in earth as in the women and the participation in the act of reproduction. Francis d'Eaubonne coined the term 'ecofeminism' in 1974. The theory of ecofeminism propelled as the combination of ideas that support the fight for women empowerment and that of sustainable environment, hand in hand. Ecofeminism seeks to bridge a connection and subsequently exploring the conceptual connections between environmental degradation and sexist oppression. Women particularly belonging to rural and poor households while performing their role as farmers or as collectors of water and fire-woods have a close connection with their local environment. Women actively participate in environmental protection movements than men as women are directly affected and influenced by degradation of nature than men. Therefore, women have been involved in several governmental & non-governmental initiatives and environment programs for the purpose of preservation, protection and conservation of natural resources.

Western and Indigenous Approaches of Ecofeminism

In order to trace the emergence and fructify of ecofeminist movement, it has linkage with diverse thoughts and practices which can be divide into two sections:

Nature-Culture Theory of the West

Having established the universal domination of nature by culture the western ecofeminists creates the foundation of a belief where women are identified with or symbolically seen to be closer to nature, as opposed to men who are associated with culture. Women are placed within the format of nature primarily because of their physiology. It is believed that her body condition her social roles and psychic (emotional, mental) structure in such a way that removes her from cultural functions by putting constraints on her.

Sherry Ortner, an American anthropologist has formulated the nature-culture approach which associates women with nature and men with culture. According to Ortner , the culture at some level of awareness asserts itself to be not only distinct from, but superior in power to, nature and in that sense the distinctiveness and superiority of man rests precisely on his ability to transform, to socialise and to culturalise nature (1972, p.11). As per this approach, women share a very close relation with nature and the resultant insubordination of both nature as well as women. Ortner draws her understanding from Simone de Beauvoir's Second Sex which highlights the fact that women are physiologically closer to nature and therefore inferior. For this Beauvoir gave example of certain female organs such and processes which are geared towards the reproduction rather than the personal needs of the woman herself.

Ariel Salleh, an Australian sociologist attempts to create a theory which stresses to include the oppressed groups of the society to become member and frontrunner of the movements related to environmental activism. Ecofeminism according to her is a potential strategy for social action that includes both men and women.

Purush-Prakriti Theory of India

Vandana Shiva, an Indian ecofeminist, as postulated by Indian philosophical thought looks at both male and female as the expression of the same person, not separate from each other. According to theory propagated by Shiva (1988), both animate and inanimate nature is an expression of Shakti, the feminine and creative principle which along with the masculine principle that is Purush creates world.

Shiva believes colonialism as a reason for destruction of nature. It was believed that advent of science and technology is responsible for breaking up of synthesis between nature, masculine and feminine principles which is thereafter replaced by an unequal and hierarchical relationship. Development proved maldevelopment for women and commencement of perpetuation and domination of nature by patriarchal control.

Shiva insists that marginalised women and communities play a very important role as they are store-house of ecological knowledge of what production and protection of life involves. The colonial practice of commercial forestry and scientific management of forest and its resources eroded the traditional indigenous knowledge to great extent and displaced the women's subsistence economy.

Bina Agarwal, prominent Indian economist finds Shiva's theory different from the Western ecofeminist perspective as theory formulated by Shiva takes into account the linkage between development and developmental changes and their subsequent impact on the environment. Shiva's taking cognisance of people's dependence on the environment for their livelihood is also appreciated.

Alternative Approach

The need to encapsulate aspects which point towards the diverse ties between women and the environment has led to the development of alternative theory which aims to be more inclusive as compared to Western and Indian ecofeminist approach.

The most prominent alternative theory is formulated by Bina Agarwal is of feminist environmentalism according to which the link between women and environment can be seen through the dynamics of gender, class, caste, race and through the organisation production, reproduction and distribution. This approach involves a perspective which calls for struggle over both resources and meaning. It involves the inclusion of ideas that include appropriation of resources by dominant groups of our society by controlling power and property. Thus, she stresses on the need to investigate the relationship between people and nature in terms of appropriation by a few.

Women – Led Environmental Activism in India

Bishnoi Movement

Long back in 1731 first trace of women movement for environment protection popped up when Bishnoi people of Rajasthan first used the Chipko tactics against tree felling. Bishnois movement began when Maharaja Abhay Singh gave royal order to cut down Khejri trees (which were worshipped by Bishnois) for construction of fortress. Under the leadership of Amrita Devi, villagers hugged the trees to stop the wood-cutters. Amrita Devi and her three daughters were beheaded for retaliation which in turn further intensified the revolt. Thus the king was forced to take back his orders.

The Bishnoi movement laid the foundation stone of women-led environment movement in India, thus the movement was revived in 1972 with the name of Chipko Andolan by Gaura Devi in Uttarakhand.

Chipko Movement

Greatest example of environmental activism in India driven by women was seen during Chipko movement (1973-75). The movement took place in 1973 in Gopeshwar in Chamoli District of northern India where ash trees of the nearby forest were to be cut down by a sports goods contractor from Allahabad. Founder of the movement Chandi Prasad Bhatt influenced the villagers of Reni to take action against the auctioning off of the forest neighbouring the village. The men of the village decided to approach the government but even after this the indiscriminate cutting of trees began. Seeing all this, the women led by a 60 year old widow Gaura Devi went ahead and hugged the trees when contractors came to cut them down foiling their attempts. Women of the village came forward in protest, hugging (Chipko in Hindi) the trees and were successful in obstructing the motive of authorities.

The Chipko movement takes on different issues for the purposes of activism such as anti-mining, timber contracting and anti-alcohol movements. All these trace their original source to the movement for environmental rights.

Someshwar Valley Movement

Another example of the kind was seen in Khirakot, a village in the Someshwar valley of Uttarakhand where women found their access to forest blocked by a miner who was planning to build a soapstone mine in that area. The dust from the mine used to settle on the agriculture fields making them difficult to plant the crops and plough the field. Women of the area therefore launched a protest against the miner. As a result miner filed criminal cases against the village people, in turn men of the village were intimidated but women were not. They in fact blocked the route to the quarry. District Magistrate of the region was moved by the struggle of village women resulting in immediate closing of the mine in 1982. This proved to be a victory for women's movement.

Bhopal Gas Tragedy

One of the remarkable instances of women's involvement in the fight against human and environmental injustice was the Bhopal gas tragedy which occurred in 1984. It caused death of more than 4000 people due to gas poisoning from Union Carbide factory and in terms of long-term signs of gas poisoning it recorded 5,00,000 victims. To unearth the information as well as to fight for proper medical care of victims women took the lead in protests. As a result, finally, the activists won rupees 360 lakh as a three-year relief grant along with government's green signal to reopen the case against Carbide factory authorities.

Narmada Bachao Andolan

Narmada Bachao Andolan was the next noteworthy environment movement. This movement challenged the building of a number of dams along the Narmada River which were funded by World Bank. As the construction of Sardar Sarovar Dam affect the environment and settlement of people of Gujarat, Madhya Pradesh and Maharashtra; people belonging to these areas stood up in revolt. These protests were led by led by the Medha Patkar, Baba Amte, and Arundhati Roy. The issue gained support from NGOs of all around the globe compelling

World Bank to withdraw this project in 1993 though the case continued in Supreme Court.

Eminent Women Environmentalists of India

- **Amrita Devi** became the pioneer in the category of women environmentalist by protesting the felling of trees on the order of Maharaja of Jodhpur. She belonged to Bishnoi community which is known for their love for nature.
- **Gaura Devi** adopted the Chipko tactics founded by Amrita Devi and organised 27 women to obstruct the cutting of trees in Chamoli district of Uttarakhand. They guarded the trees whole night until the contractors surrendered and left the village. Subsequently, news of the protest soon spread to neighbouring villages and same acts were repeated in other parts of Uttarakhand whenever some capitalist tries to tamper their forest resources. Thus women were seen as providing environmental solutions.
- **Sugatha Kumari**'s name is noteworthy as a poet and environmentalist who played a path breaking role in the protests of Silent Valley Movement. It is when the Kerala government was adamant about building of hydro-electric dam in the Palakkad district, she organised a protest to protect the evergreen tropical forests of the area and named her organisation Prakriti Samrakshana Samithi.
- **Sunita Narain**, the Director General of Centre for Science and Environment was also a member of task force for conserving tigers in Sariska (in 2005). She is considered a prominent environmentalist whose field of interest is global democracy and local democracy as well. She has great contributions regarding the issues of climate change and conservation of forest and water resources.
- **Maneka Gandhi**, a famous politician, environmentalist and animals rights leader founded an organisation in 1994 named People for Animals. She worked for the welfare of animals and to stop cruelty against them. She is now chairing the Jury of International Energy Globe Foundation which presents awards to the best environmental innovations of the year.

- **Vandana Shiva** is a Delhi based environmentalist as well as eco-feminist. She has done remarkable work in protecting forests, conserving biodiversity and organising women's networks. For this, she has written many books like Monocultures of the Mind, Staying Alive: Women, Ecology and Development, Biopiracy and Soil Not Oil: Environmental Justice in an Age of Climate Crisis. She is founder of Navdanya (1991), an organisation set up to protect the diversity of living resources. Under this initiative, approximately 2000 varieties of rice have been conserved and 34 seed banks have been established nationwide.
- **Radha Bhatt**, an activist in Nadi Bachao Andolan of Uttarakhand in 2008 opposed the construction of hydel power projects which threatened the flow of Ganga and its tributaries. These projects can also imperil the delicate forest ecosystem of the region. For this, she led a 2000km march to spread the words of water rights among the people.

Way Forward and Importance of Ecofeminism in Contemporary Times

The need of the hour is to formulate a theory which becomes an inspiration for a movement uniting the fight for the protection and conservation of environment, and women's rights. Ecofeminism comes as a theory of such kind which unites rather than alienates women and men across the world.

Most of the environmental movements undertook by women seems to compliment both fight of women for their own rights as well as to the environment. Most of the women's organizations established for environmental activism work for the cause of protection of women also such as Mahila Mangal Dals established during Chipko movement. Later on, they worked for the cause of anti-mining, timber contracting and anti-alcoholism. In rural areas alcohol addiction is directly related to domestic violence against women. Therefore, the movement can be seen necessary for protecting both environment and women.

Involvement of female participants in environmental activism acts as a catalyst in the process of environment conservation. One such example can be seen in the Bishnoi people as they are still upholding environment friendly principles such as saving water by traditional water harvesting system, indigenous cultivation method and not killing any animals (after being inspired by the heroism and determination of Amrita Devi).

As women are closer to nature they are always able to realise environmental issues better. Since women are the prospective users of the facilities provided by mother nature, it is necessary to consider their views in planning to save the environment and implementing projects for the purpose. If done so environment conservation can touch great heights. This is shown by Vanadana Shiva who under the aegis of her NGO Navdanya practices organic farming and many biodiversity conservation projects.

Conclusion

In the current scenario of global crisis vis-a-vis the environment, we must strive to protect it. Historically, culturally, physiologically and through institutional mechanisms women are proved intrinsically linked with the environment and nature. Moreover, women play an essential role in the management and appropriation of resources such as soil, water, forests and energy and have a profound traditional and contemporary knowledge of the natural world around them. This linkage is seen in the suppression and exploitation of both also. In this view, environmental movements spearheaded primarily by women for the protection and sustainable development always turned out to be successful and path-breaking. Therefore, ecofeminism emerged as an ideology that aims to bring together the shared causes of women as well as environment.

While on one hand Western ecofeminist correlates women with nature and men with culture and put forward masculinising of women for their emancipation, the Indian ecofeminist thinkers relates women with nature and men with purush (person) and blamed colonial forces and discriminate distribution of resources as the main reason for subjugation and exploitation of women and nature both.

Great applause of women-led environment activism is the common thread for all the school of thoughts among the different theorists of ecofeminism. In contemporary times, women can act as catalyst force to create awareness regarding environment concerns.

References

Agarwal, Bina (2007). 'The gender and environment debate: Lessons from India.' In Mahesh Rangarajan (Eds.). Environmental Issues in India: A Reader. New Delhi: Pearson India.

Akwa L.et al (2008). Analysis of Fuel Wood Utilization Among Rural Women in Akwanga Area of Nasarawa State. Nigeria. The Abuja Journal of Geography and Development. Vol. 1(2).

Bhatt, C.P. (1992). 'Chipko Andolan. Forest Conservation Based on People's Power', Environment and Urbanization, 2(1): 7-16.

Braidotti, R., E. Charkiewicz, S. Hausler, S. Wieringa (1994). (Eds.). Women, the Environment and Sustainable Development: Towards a Theoretical Analysis. London: Zed Books with INSTRAW.

Kumar, Radha (1993). The History of Doing. New Delhi: Kali for Women.

Mahesh Rangarajan (Eds.). Environmental Issues in India: A Reader. New Delhi: Pearson Longman.

Navdanya. http://www.navdanya.org/home

Ortner, Sherry, B. (1972). 'Is female to male as nature is to culture?'. Feminist Studies, 1(2), 5-31.

Rao, Manisha (2012).'Ecofeminism at the Crossroads in India: A Review', DEP, 20(12): 124- 142.

Salleh, Ariel (1993). 'Class, race and gender discourse in the ecofeminism/ deep ecology debate'. Environmental Ethics, 15, 225-44.

Sarin, Madhu (2001). 'Empowerment and disempowerment of forest women in Uttarakhand, India'. Gender Technology and Development. 5(3).

Shiva, Vandana (1988). Staying Alive: Women, Ecology and Survival in India. New Delhi: Kali for Women.

Shiva, Vandana, Barker, Debbie & Lokhart, Caroline (2011). The GMO Emperor has no Clothe. A Global Citizens Report on the State of GMOs - False Promises, Failed Technologies. Synthesis Report Published by Navdanya International.

Warren, Karen & N. Erkal (1997). Ecofeminism: Women, Culture and Nature.

Warren, Karen (2000). Ecofeminist Philosophy: A Western Perspective on What it is and Why it Matters. Rowman and Littlefield.

20

Estimation of Ground Water Resources of Chandernagore Municipal Corporation and Hugli-Chinsurah Municipality: A Comparative Analysis

Dr. Malay Ganguli

Assistant Professor in Geography (W.B.E.S.), Sister Nibedita Govt. General Degree College for Girls, Kolkata, West Bengal, India

Abstract

India, with 20 per cent of the world's population, has to cope with 4 per cent of its freshwater resource (Black, 2005). More demand for water leads to the over exploitation of water. Over exploitation of ground water sometimes leads to deterioration of environment as excessive rate of groundwater depletion triggers a series of environmental problems such as subsidence, compaction of aquifer, deterioration in quality of water etc. The less recharge of groundwater due to the increasing proportion of built-up area along with urbanization aggravates the problems. In India, the problem of water availability with adequate quantity and quality is being observed in major urban units but in due course of time the problem is also getting into the small urban units of the country too. In India, the study on ground water resource has got priorities recently and most of the work has been done in the metro cities. As a result, small cities are out of this type of investigation. A few research and government reports only highlight the conditions of the small cities in India. The present research work is an attempt to look into detail study of ground water resources in Chandernagore Municipal Corporation (CMC) and in Hooghly-Chinsurah Municipality (HCM).

Keywords: Ground Water, Static Water Table, Storage, Specific Yield, Over-Exploitation.

Introduction

Water lying beneath the surface of the earth is divided into two zones broadly, saturated and unsaturated zones. The upper most level of saturated zone is called as ground water level and the water consisted in this zone is called ground water. This level is controlled by hydro-geological set up and human interference. Water in the saturated zone is the only underground water that is available to supply wells and springs and is the only water to which the name ground water is correctly applied to (Heath, 2004). Ground water is the largest source of fresh water on the planet (Raghunath, 2006). Ground water system is much complicated in alluvium than that of the hard rock hydro-geological set up. In alluvium, ground water mostly belongs to leaky aquifer system. Clay layer makes the division between the upper and lower aquifer system (Mathur, 1969; Sing, 1983). Clay behaves like an aquitard in between the vertical sequence of aquifer (Datta, 2007). The fluctuation of ground water refers to the difference between the pre-monsoon and post-monsoon water levels. This helps to estimate the dynamic ground water resource of an area. This fluctuation is determined by rainfall and the tube-well pumping of the area (Mathur, 1969). Withdrawal of ground water is increasing putting a heavy burden on the aquifer system. The pressure on aquifer is getting increased in due course of time (Shankar, Kulkarn & Krishnan, 2011). After crossing the limit of natural availability withdrawal leads to fall of ground water level. Over-exploitation of ground water results into decline of ground water level which reduces the longevity of wells and tube wells (Ram & Bilas, 2010). Over-exploitation also leads to higher cost of pumping (Ram & Bilas, 2010).

The present research work is an attempt to look into detail study of ground water resources in Chandernagore Municipal Corporation (CMC) and in Hooghly-Chinsurah Municipality (HCM) of Hooghly district, West Bengal. Most of the work on assessment of ground water resource has been made in a basin area; but recently few works have been done on ground water resources mainly in metro cities. However, the present work has been done in two small adjacent cities which are located on newer alluvium. In alluvial terrain the ground water divide does not coincide with the watershed boundary which is found in the case of hard rock terrain (CGWB, 2009). Therefore, CMC and HCM which are located on the alluvial terrain have been chosen as ground water resource assessment units.

Database and Methodology

The present research work is based on both primary and secondary database. Static water table data (1995 - 2004) has been collected from State Water Investigation Directorate, West Bengal. A few information regarding static water table has also been used from various reports of the concerned cities. Characteristics of aquifer of the study area have been identified with the help of lithological logs which are collected from the water supply departments of CMC and HCM. The information related to past water supply system of the cities has been gathered from different literatures such as different reports and Draft Development Plans (DDP, 2007 - 2012) etc. Linear regression model has been used to find out trend of static water table for both the cities.

Table 1: Detailed Sources of Primary and Secondary Database

Data/ Information	Collected from	Year
Static water table	SWID, West Bengal	1995-2004
Population data	Census of India	2011
Water drafting data	Water supply department of CMC and HCM	2011
Lithological logs	Water supply department of CMC and HCM	2011
Normal meteorological data	Rice research station, Chinsurah	2008
Ward maps	Concerned cities	2009
Point positioning data	Garmin GPS	2010-2012

Source: Compiled by author.

Ground water resource of the study area has been estimated by the methodology provided by Ground Water Estimation Committee (GEC, 1997). In a few cases the equation of GEC has been modified to some extent to get better response and also because of availability of inadequate database. The used equations are as follow:

Annual recharged water (cumec) = Area (sq. m) *water table fluctuation for pre-monsoon and post-monsoon (m)*specific yield (fraction value)

Stage of ground water development

= (Current annual gross ground water draft (ham) / Net annual ground water availability (ham)) * 100

Ground Water Resources of CMC and HCM

Fluctuation of water table and specific yield of the geological materials in which the fluctuation occurs, determine the amount of recharged ground water of a particular geographical area for a year. This amount of water resource is called dynamic ground water resource of the area. Clay is the dominant geological material on the top most layer of the area. The average thickness of clay from the surface is 27.5 m. The maximum and minimum depth of static water table in pre-monsoon and post-monsoon belong to clay layer. The average fluctuation of water table in CMC was 3.53m in 1996 which reduced to 2.76m in 2004. The average fluctuation of water table in HCM was 4.21 m in 1996 which reduced to 3.74 m in 2004. Though the relationship between the year and water table fluctuation is not so much prominent, yet the reduction in fluctuation of water table in 2004 in comparison to 1996 states that the amount of recharged water from all sources has reduced. Mathur (1969) has mentioned that the fluctuation of ground water level is mainly controlled by rainfall and tube-well pumping of the area. The amount of recharged ground water was 466.5954 ham and 436.2402 ham for CMC and HCM in the year of 1996 which reduced to 364.8168 ham for CMC and 387.5388 ham for HCM respectively in the year of 2004. Therefore, the reduction in the volume of recharged ground water is 21.81 per cent in CMC and 11.16 per cent in HCM since 1996 to 2004, though the amount of recharged ground water is very much fluctuating over the years. The dynamic ground water resource of the area has been estimated using the fluctuation of pre-monsoon and post-monsoon water table of 2004. The amount of recharged ground water of CMC and HCM were 364.8168 ham and 387.5388 ham respectively.

The amount of ground water withdrawal of CMC was 645.3200 ham in 2005 and it increased to 1067.6250 ham in the year of 2011. Therefore, the amount of ground water withdrawal has increased almost 65 per cent in 2011 from 2005. The amount of ground water withdrawal of HCM was 725.9850 ham in 2005 which increased to 1595.1138 ham in 2011. Therefore, the amount of withdrawal has increased almost 120 per cent in 2011 from 2005. Total ground water discharge in CMC was 645.3200 ham in 2005 which raised to 1095.26 ham in 2011(1067.6250 ham by corporation + 27.6350 ham by apartments) and also the figure of total ground water discharge of HCM has changed drastically from 725.9850 ham in 2005 to 1614.9974 ham in 2011 (1595.1138 ham by municipality + 19.8836 ham by apartments). Net annual ground water availability was 364.8168 ham in CMC and 387.5388 ham in HCM in 2004. This annual availability of ground water follows a declining trend, whereas the withdrawal of this resource follows highly accelerating trend. Stage of ground water development explains the original picture of a place in terms of recharge and discharge of water. Stage of ground water development of CMC was 176.88 per cent in 2005 which increased to 300.22 per cent in 2011 and the stage of ground water development of HCM was 187.33 per cent in 2005 which increased to 416.73 per cent in 2011 (the stage of ground water development has been calculated on the basis of net availability of ground water of 2004 for both the cities). The above figure states that the amount of discharge increased from 1.7 times in 2005 to 3.0 times in 2011 of recharge in CMC. The amount of discharge increased from 1.8 times in 2005 to 4.1 times in 2011 of recharge (Table - 2). Therefore, both the cities come under *over-exploited zone*.

Table 2: Estimated Ground Water Resources

Attributes	CMC	HCM
Area (sq.km)	22.03	17.27
Water level fluctuation (m), 2004	2.76	3.74
Recharge (ham), 2004	364.8168	387.5388
Water withdrawal (ham), 2005	645.3200	725.9850
Water withdrawal (ham), 2011	1067.6250	1595.1138
Water withdrawal in apartments (ham), 2011	27.6350	19.8836
Total ground water discharge (ham), 2011	1095.26	1614.9974

Attributes	CMC	HCM
Net annual ground water availability (ham), 2004	364.8168	387.5388
Net annual ground water availability (LPCD), 2011	59.86	59.91
Stage of ground water development (%), 2005	176.88	187.33
Stage of ground water development (%), 2011	300.22	416.73

Source: Computed by author.

Storage

Ground water is stored in the pore spaces of geological materials. The amount of water stored in underground stratum is called storage. Water belongs to the annual fluctuation zone of a particular year and is termed as dynamic ground water resource. However, water resource lies beneath the dynamic ground water resource zone is called static ground water resource. Decline of static water table leads to exploitation of static ground water resource of an area. CMC and HCM cities are situated on the newer alluvium of Holocene age. There is a good opportunity provided by the nature to be a significant ground water storage zone. Irrespective of this, it is observed that we are losing this unique resource with huge amount annually. The total loss of ground water storage is 664.8654 ham in CMC and 675.6024 ham in HCM since 1995 to 2011 (Table - 3). The average annual storage loss is maximum during 2004 to 2011 in both the cities because of huge withdrawal of ground water in both the cities (Figure - 2.10). The average annual storage loss is observed 111.6825 ham in CMC and 111.8651 ham in HCM during 1995 to 2011.

Table 3: Ground Water Storage Loss in CMC and HCM during 1995-2011

Periods	CMC (ham/ period)	HCM (ham/ period)	CMC (ham/ year)	HCM (ham/ year)
1995-2000	76.6644	117.0906	15.3328	23.4181
2000-2004	114.9966	80.8236	28.7491	20.2059
2004-2011	473.2044	477.6882	67.6006	68.2411
Total	664.8654	675.6024	111.6825	111.8651

Source: Computed by author based on pre-monsoon static water table data.

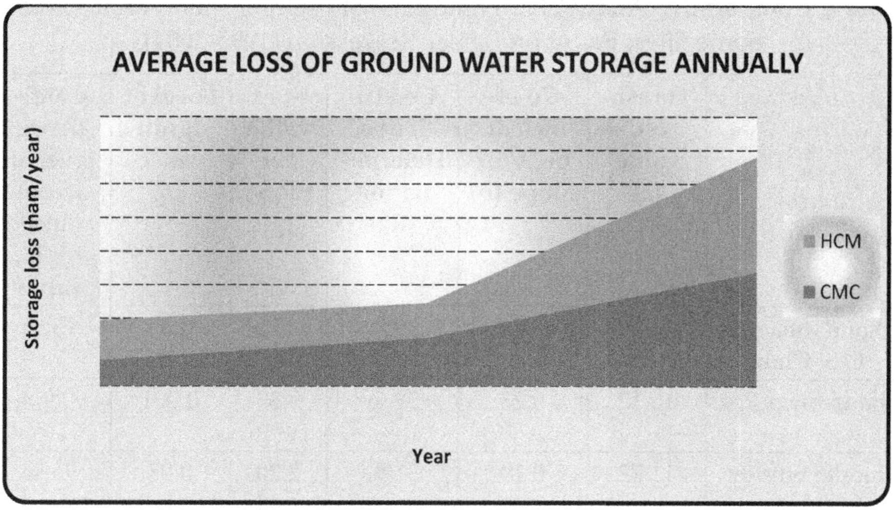

Figure 1: Average Loss of Ground Water Storage Annually

Water Level

The analysis of water level of the study area has been done based on four observatory wells. Water level is different in different sites. All the water levels have been recorded from ground surface. Static water table or static water level refers to the level of water in a well when the pump is not operating (http://www.waterbank.com). From the linear regression analysis of pre-monsoon static water table of four observation sites it is found that the threshold value (a= distance between the point of origin and the point where trend line touches the Y-axis) is maximum at Chandernagore O.S. Club (a= 13.90) and minimum at Adisaptagram B.S.F. (a=8.14). The 'a' value shows that the depth of static water table is maximum at Chandernagore O.S. Club and minimum at Adisaptagram B.S.F. The co-efficient of 'X' or slope value (b) is maximum at Bhadreshwar P.S. (b=0.23) and minimum at Adisaptagram B.S.F. (b=0.12). The 'b' value shows the gradient of the falling water table from 1995 to 2004. The falling rate is maximum at Bhadreshwar P.S. and minimum at Adisaptagram B.S.F (Table - 4; Figure - 2).

Table 4: Comparative Analysis of Trend Line of Pre-monsoon Water Table of Some Sites, Based on Linear Regression (1995-2004)

Sites	Thresh-old value (a)	Co-ef-ficient of 'X' or Slope (b)	Co-effi-cient of Determi-nation (R2)	t-test value for Slope (b)	Level of signifi-cance	Confi-dence level of accep-tance of 'b' value
Chandannagore O.S. Club	13.90	0.21	39 %	2.25 (v=9)	0.05	95 %
Bhadreswar P.S.	10.12	0.23	58 %	3.35 (v=9)	0.001	99.9 %
Bandel Survey T. Centre	11.72	0.19	31 %	2.20 (v=8)	0.05	95 %
Adisaptagram B.S.F.	8.14	0.12	14 %	1.25 (v=9)	0.1	90 %

Source: Computed by author.

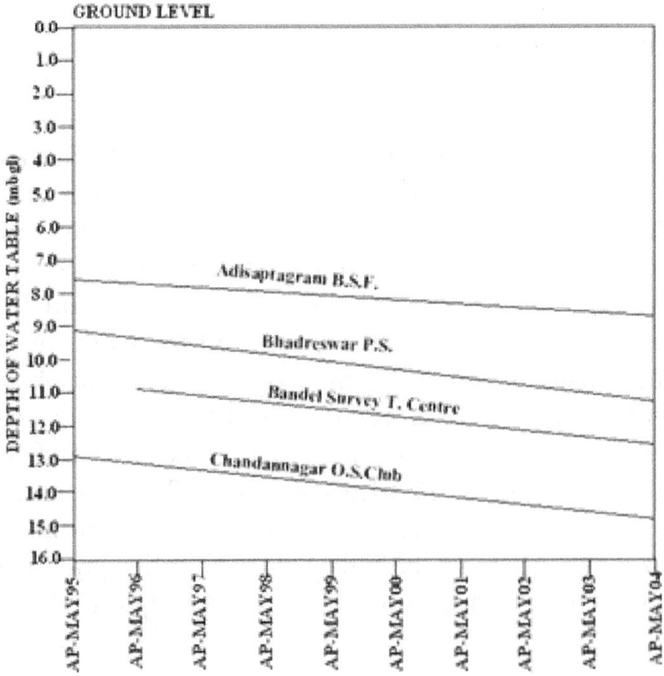

Figure 2: Trend Line of Pre-Monsoon Water Table

From the pre-monsoon water table of 1995 to 2011 it is observed that water table declined from 11.50 mbgl to 16.53 mbgl in CMC and 8.57 mbgl to 15.09 mbgl in HCM during 1995 to 2011 (Figure - 5). Therefore, the average falling rate of pre-monsoon static water table is 0.31 m/year in CMC and is 0.40 m/year in HCM. The pre-monsoon static water tables of 1996 (Figure - 5) and 2004 (Figure - 6) show that the deepest part of water table belongs to Central-Eastern part of CMC for the whole area covering both the cities.

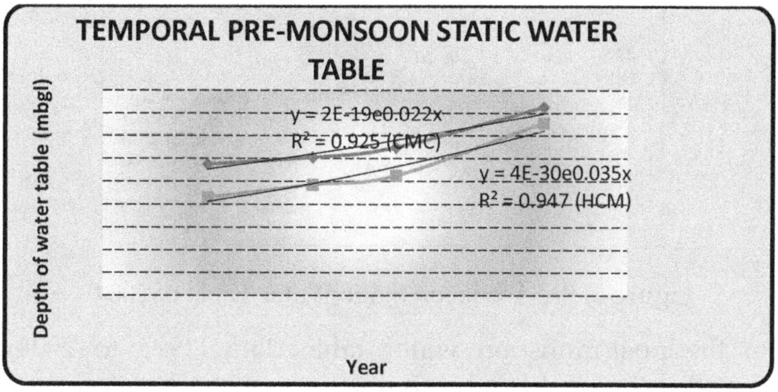

Figure 3: Temporal Pre-Monsoon Static Water Table

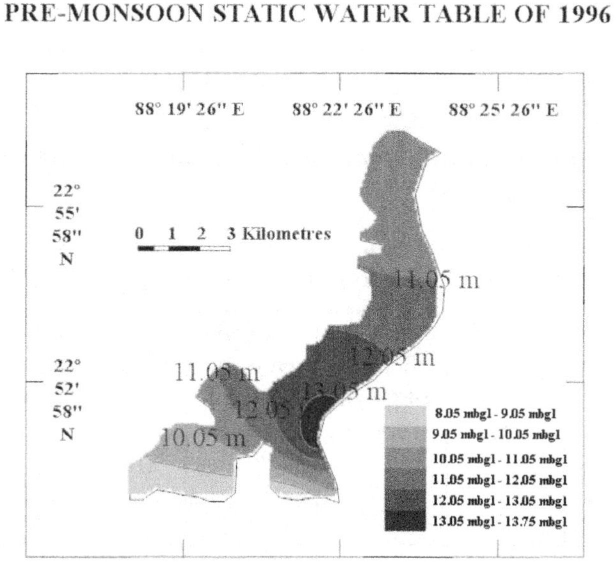

Figure 4: Pre-Monsoon Static Water Table of 1996

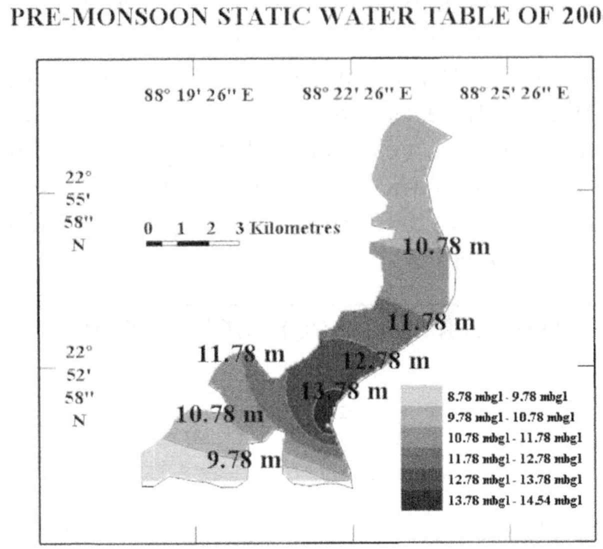

Figure 5: Pre-Monsoon Static Water Table of 2004

From the post-monsoon water table data (1995 to 2004), it is found that the threshold value is maximum at Chandernagore O.S. Club (a=10.14) and minimum at Adisaptagram B.S.F. (a=3.51). The 'a' value indicates that the depth of static water table is maximum at Chandernagore O.S. Club and minimum at Adisaptagram B.S.F. The co-efficient of 'X' or slope (b) is maximum at ChandernagoreO.S.Club (b=0.23) and minimum at Adisaptagram B.S.F. (b=0.10) which shows that the rate of falling of water table is maximum at Chandernagore O.S Club and minimum at Adisaptagram B.S.F. during the period (1995 to 2004) (Table - 3; Figure - 8).

Table 5: Comparative Analysis of Trend Line of Post-monsoon Water Table of Some Sites, Based on Linear Regression (1995-2004)

Sites	Threshold value (a)	Co-efficient of 'X' or Slope (b)	Co-efficient of Determination (R2)	t-test value for Slope (b)	Level of significance	Confidence level of acceptance of 'b' value
Chandannagore O.S. Club	10.14	0.23	55 %	3.15 (v=9)	0.01	99 %

Sites	Thresh-old value (a)	Co-ef-ficient of 'X' or Slope (b)	Co-effi-cient of Determi-nation (R2)	t-test value for Slope (b)	Level of signifi-cance	Confi-dence level of accep-tance of 'b' value
Bhadreswar P.S.	6.91	0.19	65 %	3.88 (v=9)	0.001	99.9 %
Bandel Survey T. Centre	7.99	0.14	33 %	1.83 (v=8)	0.1	90 %
Adisaptagram B.S.F.	3.51	0.10	25 %	1.61 (v=9)	0.1	90 %

Source: Computed by author

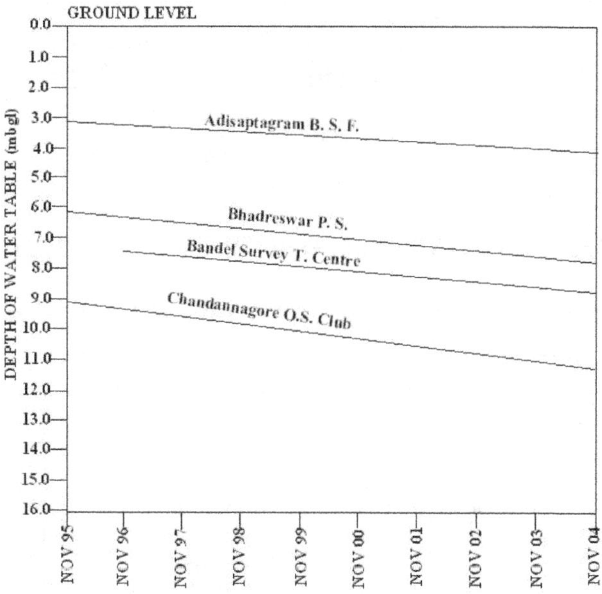

Figure 6: Trend Line of Post-Monsoon Water Table

From the slope analysis of pre-monsoon and post-monsoon static water table (1995 to 2004), it is found that the gap between the gradient of pre-monsoon and post-monsoon water table is maximum at Bandel Survey T. Centre (pre-monsoon: b=0.19, post-monsoon: b=0.14) and is minimum at Chandernagore O.S. Club (pre-monsoon: b= 0.21, post-monsoon: b=0.23) and Adisaptagram B.S.F. (pre-monsoon: b=0.12, post-monsoon: b=0.10). Irrespective of the same gap between the two sites, the falling rate of the static water table is higher at Chandernagore O.S.Club. The post-monsoon static water tables of 1996 (Figure - 9) and 2004 (Figure - 10) show that the deepest part of water table belong to Central-Eastern part of CMC for the whole area covering both the cities.

POST-MONSOON STATIC WATER TABLE OF 1996

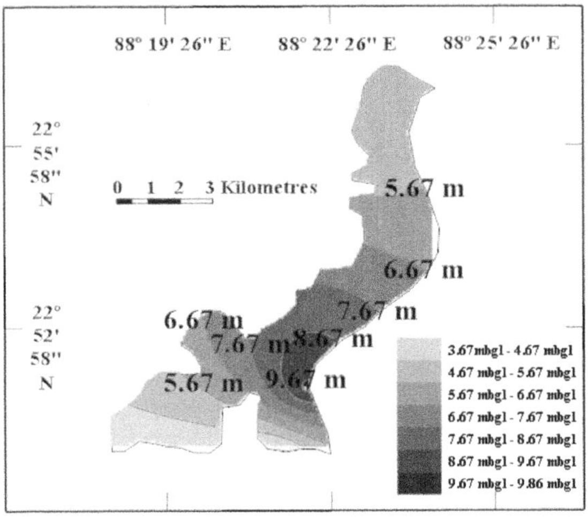

Figure 7: Post-Monsoon Static Water Table of 1996

POST-MONSOON STATIC WATER TABLE OF 2004

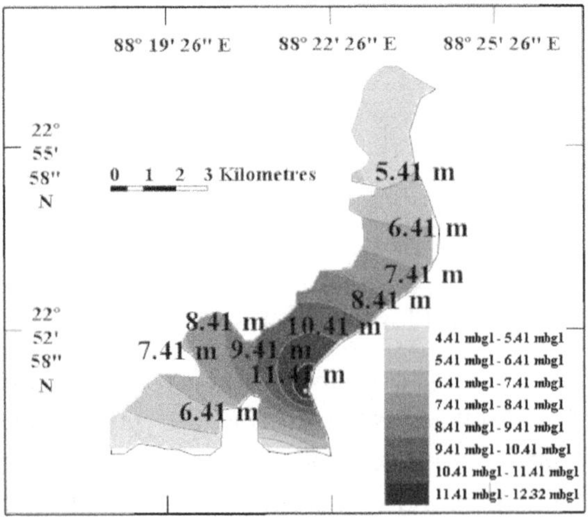

Figure 8: Post-Monsoon Static Water Table of 2004

Conclusion

The main issues of the cities are almost same as they are located in the same physical and socio-cultural environments. Both the cities come under over-exploited zones and it is estimated that the magnitude of ground water depletion has been enhanced to a high degree in comparison to the past. Therefore, static water table is declining rapidly. This induces compaction of geological materials and land subsidence. Moreover, this rapid decline of static water table and spacing of pump houses in an unscientific manner have reduced the average age of pump houses as well as their yielding capacity from the past.

References

Black, M. (2005). *The no-nonsense guide to water.* Jaipur: Rawat Publications.

Central Ground Water Board. (2007). *Hydrogeology and ground water resources of Hugli district.* Kolkata: S. Datta.

Central Ground Water Board. (2009). *Detailed guidelines for implementing the ground water estimation methodology.* New Delhi: Ministry of Water Resources.

Draft Development Plan (2007/08-2011/12): Hooghly-chinsurah municipality (Unpublished document).

Draft Development Plan (2007-2012): Chandernagore municipal corporation (Unpublished document).

Heath, R.C. (2004). *Basic ground water hydrology.* USGS: USGS Water-Supply Paper. Retrieved March 2010, from http://www.usgs.gov/

Mathur, R. N. (1969). *A study in the ground water hydrology of the Meerut district, U. P., India.* Varanasi: Banaras Hindu University.

Raghunath, H. M. (2008). *Ground water.* New Delhi: New Age International (P) Limited.

Ram, R. & Bilas, R. (2010). Groundwater management in Chandauli district, U. P.: A case study, *National Geographical Journal of India,* 56, 93-100.

Shankar, P. S. V., Kulkurni, H. & Krishnan, S. (2011, January 8). Indian's groundwater challenge and the way forward. *Economic & Political Weekly,* pp. 37-44.

Sing, U. (1983). *Assessment and prospects of water resources in Eastern U. P.* Banaras: The National Geographical Society of India.

21

A Perceptive Study on Tourism Practices and Environmental Quality with Respect to Tourism Footprints in Rishyap Village of Kalimpong District, West Bengal

Dr. Anirban Roy[1] & Partha Das[2]

[1]Assistant Professor and Head, Department of Geography & Environmental Science, Government General Degree College, Paschim Medinipur, West Bengal, India

[2]Assistant Professor, Department of Geography, A. B. N. Seal College, Cooch Behar, West Bengal, India

Abstract

Tourism is one of the fastest growing economic sectors in India and is one of the major pillars of economy in North Bengal. Volume of tourists and tourism economy, both have grown many folds here during the last decade. In recent years North Bengal has witnessed booming growth in tourism infrastructures, certainly resulting into greater economic dividend for the local people as well as exerting considerable pressure on the environment in such ecologically sensitive landscape. Thus, a conflict of interests has come up with economic dividends being on the opposite end of environmental needs. Now whether this trend of tourism growth is a boon or a bane for the region can be best decided by the people living there, so their perception is of utmost importance. Again, to measure the impact

[2]Corresponding Author

of tourism, concept of ecological footprint is considered where the rate of resource depletion is measured and compared against the actual yield of resources there. Thus, the exact level of stress on the environment resulting from tourism practices can be identified.

The study area, Rishyap, is a famous hill station, located in the Kalimpong district in Northern West Bengal. With its varied offerings of natural wonders, Rishyap attracts tourists throughout the year which has instigated rapid growth of tourism infrastructure lately.

But this growth is skewed towards 'popular tourism' instead of being ecologically sensible 'low key tourism'. So, this paper attempts to identify the imprints of tourism on local environment and perception of the inhabitants of Rishyap about the changing scenario taking place.

Keywords: Popular Tourism, Low Key Tourism, Tourism Footprint, People's Perception.

Introduction

West Bengal is the only state in India with snow-clad Himalayan peaks as its crown in the north, waves of Bay of Bengal washing its feet down the south and varieties of lush greeneries decorating it in almost every direction. Apart from the natural beauty, the state offers religious and heritage spots to visit also. So, West Bengal is one of the leading states in the country in terms of tourist footfall, particularly foreign tourist arrivals. The provisional annual tourist footfall reached 87.3 million in 2018 from only 14.5 million in 2005. Around 1.61 million tourists from abroad visited the state in 2018 which is around 15.27% of total foreign tourist arrival in the country in that year (India Brand Equity Foundation, 2020). The government of West Bengal has developed six tourism circuits to improve and encourage tourism in the state, out of which two are from North Bengal itself. These two are 'Darjeeling Himalayas' located mainly in the Darjeeling district and 'Dooars Foothills' spread over Jalpaiguri and Alipurduar districts. The rate of growth in tourist arrival in these two circuits has been found much more than any other circuits in the state. Volume of tourists has increased five times during the last decade (Tourism Survey of Bengal, 2014), so has been the growth in the tourism-based economy which has witnessed almost three times growth during the same period (ibid). Darjeeling hills topped the chart of top 10 tourist destinations in this state (Annual Final Report of Tourism Survey, West Bengal 2014-15, Ministry of Tourism, Govt. of India)

Rishyap, as described in the website of Dept. of Tourism, Govt. of West Bengal "is a tourist village in the Lava-Lolegaon circuit in Kalimpong sub-division". Once a secluded hill hamlet only accessible by a trek route stretching 5 km. from nearest tourist centre of Lava, Rishyap is now connected via all-weather roads to other important parts of the region. This has lately resulted in rapid increase in tourist footfall here, leading to ever-increasing accommodation facilities. Located within the Neora Valley National Park, Rishyap is an ecologically sensitive spot with a vast array of floral and faunal diversity. So, the on-going tourism practices, which may be considered beneficial for the local economy, are actually escalating pressure on the natural environment. Thus, economic growth and environmental quality are at loggerheads in Rishyap and it's when the perception of the local inhabitants becomes very important. They are the best judge of the things going around in their place, of the tourism footprint i.e. the changes or pressure brought in by tourism and the future kind of development in Rishyap.

On the other hand, tourism footprint measure is a quantitative tool for assessing the impact of tourism activities on the ecosystem of a tourist destination. Here the 'tourism ecological footprint', the 'tourism carbon footprint' and the 'tourism water or waste footprints' are measured separately using some well accepted indicators. These footprint results are very useful for local stakeholders including the service providers to increase their awareness on sustainability issues that otherwise wouldn't have been understood. So, research on tourism footprint in India, where tourism industry mainly thrives on popular tourism, is an important requirement. Tourism is considered to be responsible for 8% of global greenhouse gas emissions. According to a report, published by The Telegraph in May'18, India has the fourth largest tourism carbon footprint in the world. That's why it is apprehended that travel destinations in India are already environmentally stressed and in the immediate need of adopting eco-friendly measures and sustainable tourism practices.

Rishyap, is located in a rich bio-diversity zone. So, it's really important to maintain the environmental quality there by reducing the tourism footprint without hampering the economic growth of the place. The present study sheds light on the above-mentioned aspects.

Literature Review

This study is divided into two parts. The first part is finding out the aspects of environment and tourism practices that have some influence on the footprint matter. Then a schedule has been prepared with adding more inquiries on selected household practices and changing scenario of the surroundings. The second part is reaching door to door to get the perception of the local people so the ground scenario can be perceived.

Now for the tourism footprint portion of the first part, an array of literature has been studied. Rees (1992) was the first to introduce the concept of Ecological Footprint and along with Wakernagel (1996) he developed the concept further. Hunter (2002) formally introduced the concept of Tourism footprint and further developed the concept in 2007.Stoeglehner and Narodoslawsky (2008) have shown how measuring ecological footprint can help in land use policy making. Galli et al. (2012) and Fang et al (2014) have shown how different single dimensional footprint measures can reflect human pressure on planet and call for research related enquiries.Wang et al (2017) summarized the progress and shortcomings till then and proposed areas for further research. Wiedmann, Ercin, Knoblauch, Ewing, Patterson, Marzouki etc. have worked on this matter in recent past. For estimating the tourism footprint; Tourism Carbon Footprint, Tourism Ecological Footprint and Tourism Water Footprint have been separately measured in works of many scholars. Different analytical frameworks for Tourism Ecological Footprint have been given by Gössling et al. (2002), Zhang and Zhang (2004) etc. Similarly, framework for measuring Tourism Carbon Footprint, over the years, have been given by Becken and Patterson (2006), Sharp et al. (2006), Sun (2014) etc. Scholars like Yang et al. (2011), Hadjikakou et al. (2013) etc. have created analytical frameworks for measuring Tourism Water Footprint. In this study, the aspect of water has been taken together with Ecological Footprint and a separate Tourism Waste Footprint has been considered for measurement by the authors as the issue of waste generation and management in the study area seemed more important.

Study Area

Rishyap or Rishop is a tourist village in Kalimpong located at 27° 06´40´´N latitude and 88°39´09´´E longitude and at an elevation of 2591 m. Its serene environment and great view of the snow-capped Himalayan mountain ranges attracts tourists from all over the places. Located within the Kolbong forest this hamlet is surrounded by Pine, Birch and Fir forest which is home to numerous birds of the Himalayan origin. There is one-hour trekking route there, through dense forest, to Tiffin Dara viewpoint that offers splendid view of the mighty Kanchanjungha mountain range. So, a varied range of tourists; from bird watchers to adventurers, students from educational institutions to vacationers, visit the place. Nearest municipal town of Kalimpong is located at a distance of 28 km from Rishyap while the nearest railway station and airport are at New Jalpaiguri and Bagdogra respectively. From these places' tourists can reach the study area by cars via Kalimpong or Gorubathan. Once located at 5 km trekking distance from Lava, Rishyap is now connected via all-weather roads. This has resulted in rapid influx of tourist footfall in this hilly hamlet.

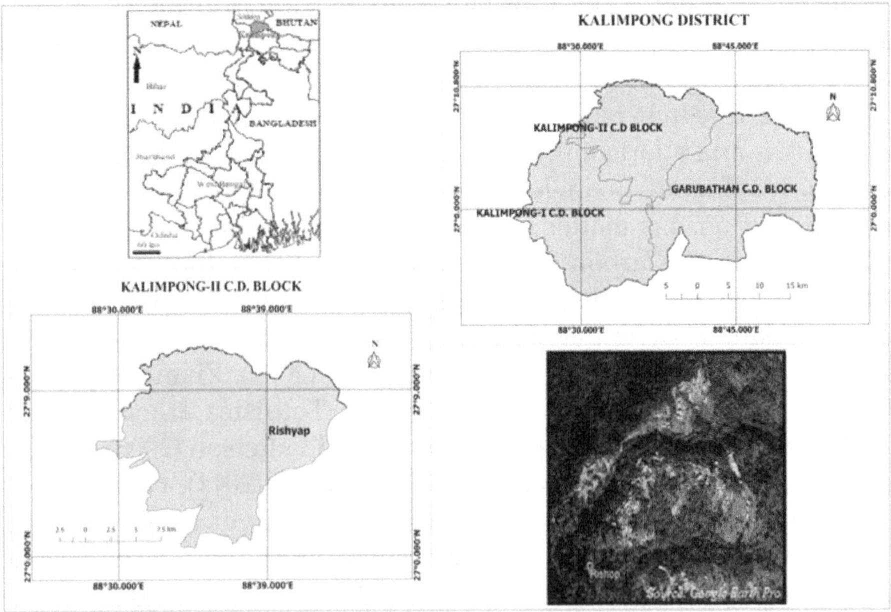

Figure 1: Location Map of the Study Area

To accommodate the growing number of tourists and grabbing the opportunity of getting economic dividend from the situation, many hotels, home stays and resorts have lately popped up here. As mentioned earlier, Rishyap being an ecologically sensitive spot, the on-going popular tourism practices and rapid growth in tourism infrastructure that seems to be beneficial for the local people for now is actually exerting unprecedented pressure on the local ecology. The tranquil environment, which is the sole USP of Rishyap is getting degraded year after year for the sake of economic benefits. On contrary, importance of this economic dividend in the life of the local inhabitants cannot be denied. So, Rishyap turns out to be an ideal spot for researching this contesting idea of tourism centric economic growth and growing environmental concerns.

Objectives

Important objectives of the present study can be stated as below-

- To bring out the present condition of tourism infrastructure and tourism practices in Rishyap village, located in Kalimpong District of West Bengal.
- To understand the behaviour of the tourists towards the place and the perception of the local people towards ongoing tourism practices.
- To delineate the impact of popular tourism on environment in relation to the growing tourism footprint there.

Materials and Methods

Present study has two separate but interlinked goals. Firstly, it tries to bring out the perception of the stakeholders on popular tourism practices as well as on the condition of the local environment and secondly, the tourism footprint resulting from the current tourism practices in the study area. Thus, the study was designed in such a way that both ends can be reconciled in a justified manner.

Secondary data have been collected from various published sources, both from the Government offices and from the online sources on tourism related infrastructure and tourist inflow, in Rishyap village. Understanding the relationship between Tourism and its footprint on environment were of utmost necessity for the preparation of survey schedule on that line and setting up the objectives of the study.

Primary data on the perception of local people on the condition of tourist facilities and local environment was collected by rigorous survey in Rishyap village with the help of a purposive Survey Schedule. A stratified-random technique has been followed throughout the survey. Entire primary survey was done in November, 2019 and a total of 108 respondents had been surveyed from among the hotels and resort operators and employees to owners of the home stays to common residents of Rishyap who belonged from 67% of the existing establishments in the study area. Later these primary data have been tabulated and array and analysis of the same have been done according to the line of the objectives taken. Software like *Google Earth Pro, MS Excel* and *MS Word* has been used for that purpose.

Tourism Footprint on environment has been calculated using both primary and secondary data. A spread sheet has been developed for this purpose. Utmost care had been taken in selecting the parameters and quotients that are most relevant for the present study.

Tourism Footprint is the summation of three separate indices i.e. Tourism Carbon Footprint, Tourism Ecological Footprint and Tourism Waste Footprint.

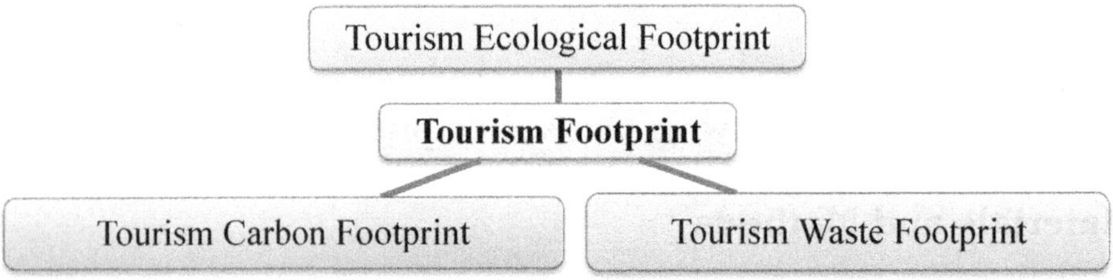

Figure 2: Different Indices for Calculation of Tourism Footprint

Tourism Carbon Footprint: TCF analysis focuses on environmental issues caused by large-scale tourism, particularly its energy consumption and CO_2 emissions. It is generally calculated by the formula

TCF = (Distance Travelled + Fuel Used) × Emission Factor

–(IBM Knowledge Centre, 2019)

For the present study, the formula has been modified and fuels used in kitchens are incorporated with the above. Thus, the equation used in present study is-

$$TCF = (\text{Distance Travelled} \div \text{Fuel efficiency}) \times \text{Emission Factor} + (F1\times EF1+F2\times EF2...Fn\times EFn) \quad ...(i)$$

[Where, F1, F2....Fn are different types of fuels used in kitchens and EF1, EF2...EFn are their respective emission factors.]

Thus,

Per Capita TCF = Total TCF/ Total no. of Tourists ...(ii)

Tourism Ecological Footprint: Tourism Ecological Footprint (TEF) measures impact of consumption by the tourists on local environment. A 'Water Component' has been added to the TEF for the present study. Thus -

TEF = Tourist Consumption Footprint (TC_1F) + Tourist Water Footprint (TWF) ...(iii)

Per Capita TEF = TEF/Total no. of tourists ...(iv)

$TC_1F = (C1 \times Y1 + C2 \times Y2....Cn \times Yn)$ ...(v)

And TWF = Total Water consumed ÷K ...(vi)

[Where C1, C2...Cn are individual products for consumption and Y1, Y2...Yn are their corresponding yield rate. K is the conversion constant]

Tourism Waste Footprint: Solid wastes produced by tourism activities are converted into emission values using different emission factors. Two types of carbon emissions have been considered here to calculate the tourism waste footprint (TW_1F) i.e. Methane (CH_4) and Carbon Di Oxide (CO_2). Further, the emission values of CH_4 have been converted into the same of the CO_2. Thus, the total carbon emission from the waste have been calculated and expressed in Global Hectare (gha).

$TW_1F = (Wt \times ECH_4) + (Wt \times ECO_2)$ ...(vii)

[Where Wt is total waste generated, ECH_4 and ECO_2 are respective emission factors.]

Per Capita $TW_1F = TW_1F \div$ Total Tourists ...(viii)

Results and Discussion

Outcomes of the present study may be discussed under the following heads:

1. Tourism; Infrastructure, Facilities and Practices

Rishyap as a tourist destination is being promoted since 1998 primarily to aid the local people economically by offering an alternative mode of earning. Initially it was accessed via a trekking route from Lava but its actual development as a tourist spot began since when it has been connected via an all-weather motor-able road and the NHPC power grid. Tourists looking for nature's bliss, peace and tranquillity started coming here to spend mostly 2-3 nights. Since then there has been a growing number of home stays, hotels and resorts coming up in this picturesque mountain hamlet. Most of the homestays are being operated by local residents while the newly built luxury hotels and resorts are leased out to private tourism operators from outside. The present study encompassed responses from all the key players of the place ranging from local residents to tourists to owners of home stays to managers and employees of the corporate run hotels and resorts. The survey found around 48% of these establishments as being built on excavated hill slopes and in case of another 11% the earlier terrain character has been changed into landscaped plots.

Rishyap, till the time of this survey, did not have any direct public transport options connecting it with the district town of Kalimpong or Siliguri, from where the residents mostly source their goods. To avail such facilities people, have to reach nearby Lava from where only one bus and one jeep commute passengers to Siliguri and back once daily. In such a scenario, tourists visiting Rishyap, till now largely depend on private or hired cars which add up to the carbon footprint of the place and an issue of parking on narrow hill roads.

Again, more than 38% of establishments in Rishyap burn wood as cooking fuel and this increases the local carbon footprint. Similarly, more than 33% establishments here use diesel run generators in absence of electricity which adds up to the carbon footprint.

Rishyap, like any other hill station, has limited supply of potable water in its surroundings and both family dwellings and tourism infrastructures largely depend on either tanker-based supply of water from Algarah water reservoir or piped water supply from nearby springs. Thus, increasing tourism in Rishyap leaves a larger water footprint in an already stressed supply scenario.

Though farming is the principal source of income for Lepcha and Sherpa people of Rishyap; they mainly produce limited amount of cereals, pulses, spices, seasonal vegetables and fruits on hillsides and some of them practice poultry farming. Though their produces are enough for personal requirement but are not sufficient to cater the demand of growing number of tourists. Rice, wheat, meat etc. are mostly outsourced in Rishyap and as around 96% of tourists ask for non-veg food items; all these add up to the ecological footprint here. Ecology here is further stressed by acts of deforestation and often by loud party music playing in tourist establishments. 25% of tourists are reported to be asking for loud sound system and around 37% of the establishments offer that.

These activities are totally unacceptable in such adjacent to forest location and add up to the ecological footprint here.

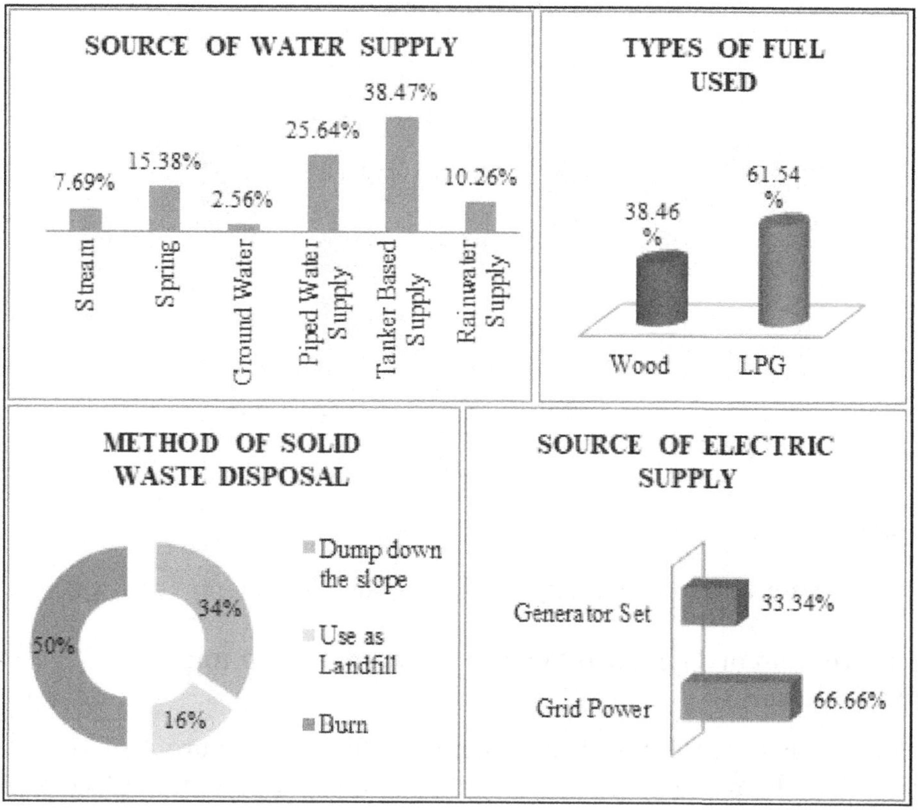

Source: Primary Survey

Figure 3: Amenities and Practices in the Survey Area

Waste disposal is a matter of serious concern in Rishyap village. In absence of any proper waste collection and disposal system, households and hotels alike mostly burn their solid waste or dump it down the hill slope. Around 83% of the establishments did not have any waste water treatment facilities, while 67% of those did not practice separating bio and non-bio degradable wastes. And with more than 65% of respondents believing that tourism in Rishyap has increased two to three folds in last decade, the tourism waste footprint leaves serious impact in the locality. Like any other places, plastic pollution is a matter of grave concern here and responsibility for that goes to both hoteliers and tourists.

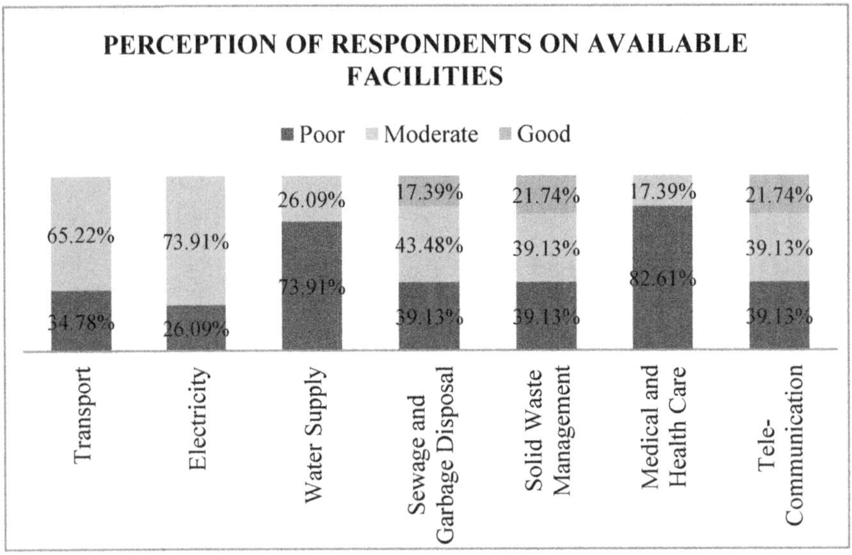

Source: Primary Survey

Figure 4: Respondents Opinion on Existing Facilities in Rishyap

2. Perception of The Local People; Tourism, Environment and Pollution

More than 68% of the respondents opined that there has been a considerable increase in the amount of built-up area in this place over the last one decade. This is evident from the fact that around 35% of the establishments surveyed showcased a sub-urban land use in an otherwise rural set up. 21% respondents apprehended that such rapid construction work has resulted into slope instability while another 15% were concerned that there has been increased runoff in the area. More than 68% of the respondents have complained of decrease in greeneries in Rishyap in last one decade, while more than 57% of them have reported decreased sightings of animals and birds during the same time period. About 30-35% of the respondents complained of increased water and land pollution in the study area, while another 26% were concerned about increased air pollution.

But this concern about deteriorating environment or increasing pollution mostly does not deter local people's enthusiasm about tourism. Around 40% of respondents did not recognize any change in morphology and hydrology due to rapid construction of tourist establishments. This is not surprising at all because around 54% of them opined that tourism is beneficial to almost all the residents of Rishyap. Though around 30% respondents complained that the benefits are almost none or limited among few stakeholders only and they also recognized tourism activities as a threat to local environment. Thus, the study shows there is a difference of opinion among the respondents regarding the impact of tourism on local environment with majority of the stakeholders having a positive attitude towards it as it brings economic benefits and better livelihood.

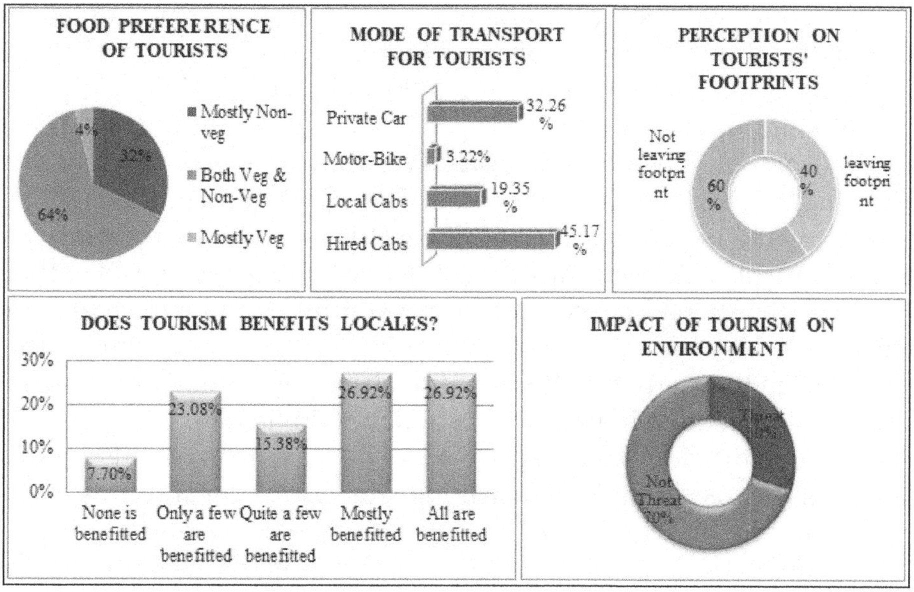

Source: Primary Survey

Figure 5: Aspects of Tourism and Respondents' Opinion Regarding it

3. Tourism Footprints in Rishyap and its Environmental Impacts

As stated before, tourism activities account for almost 8percent of global greenhouse gas emissions at present. It may also be noted that tourisms global carbon footprint has increased from 3.9 to 4.5 gt CO_2 just between 2009 and 2013 (Lenzen et. al, 2018). Transport, shopping and food are identified as the most significant contributors for this increased carbon footprint (ibid). Remote places like Rishyap, where almost each and every consumable article is brought from somewhere else, it is very likely to get a high carbon footprint. The tourism carbon footprint (TCF) in Rishyap reveals to be as high as 8.46 gt CO_2 per capita, which is almost double the world standard. This can be an effect of using low energy efficient fuels like wood, coal and diesel in the kitchens by the hotels and resorts. The study reveals that the hotels and resorts in Rishyap consumed 584 tons of coal and 2461 litters of diesel in one month during the peak season. Besides, most of the vehicles plying to and fro Rishyap uses diesel as fuel which again have high emission value. High carbon footprint directly impacts the places ambient air quality and has adverse effects on global environment.

The world average per capita ecological footprint was 2.75 gha in 2016. India has an average ecological footprint of 1.16 GHa per person and holds 164thposition in the world (Global Footprint Network, 2016). Present study reveals that, Rishyap has an ecological footprint (TEF) of 2.07 GHa per capita that is a bit lower than the world average but much higher than the same for India. The significant contributors for the high value of ecological footprint in Rishyap can be identified as the consumables like rice, meat, sugar, tea, wheat, potatoes and bottled water. As mentioned earlier, a water footprint component had been implemented with the ecological footprint for the present study. It is evident from the study that Rishyap has an exorbitant daily water draft of 87,040 litters, quite high for a small mountain hamlet, while most of it goes down the slope as the waste water generation is equally high at 84067 litters. It is certainly an alarming situation as most of this water is withdrawn from the underground water table and local surface water sources are very limited. The per capita tourism water footprint in Rishyap is calculated at 2.02 CuM per day when the same for the world is 3.40 CuM per day (Hoekstra and Champagain, 2006 and Yang M et.al, 2007).

Tourism waste footprint (TWF) is another important aspect of the present study that reveals the magnitude of pressure exerted by the popular tourism on the local environment. It is evident that the total solid waste generation is alarmingly high in Rishyap village as it accumulates 889.92 kg per day. Absence of institutional solid waste management practices are instrumental for the dumping of solid waste over the mountain slope or, what most hotels and resorts do in Rishyop, burn these to reduce the volume. Although per capita solid waste generation in Rishyap revealed to be 0.10 kg per day which is way below the world average of 0.74 kgs (World Bank, 2016).

Total tourism footprint in Rishyap village thus computed, by summing up different footprint indices together, and the value generated is 2.071532 GHa. It may be stated that the value is below the world average ecological footprint but higher than the Indian standard. So, it is the high time for Rishyap village to reshape their tourism practises towards sustainable and ecologically viable ones rather than the prevailing popular tourism generating high footprint.

Findings and Suggestions

Based on the above discussion, major findings of the study along with some suggestions to the policy makers may be listed below.

- Tourism Footprint is high in Rishyap. Although it is not higher than the world average but this is probably the right time to raise the concern. At the same time, it should be reduced without hampering the local economy.
- Low foot print tourism must be encouraged in Rishyap instead of popular tourism. That may play the key role in transforming tourism practices without hampering the economic interests of the stakeholders.
- Awareness among the tourists about the vulnerability of the local environment must be increased through extensive campaigning. Green taxes or environmental cess on the goods consumed may be a viable way to do so.
- Hotel owners should be encouraged for opting eco-friendly hospitality practices. Tourists should be served local products as much as possible instead of packaged foods or beverages that have greater ecological and carbon footprint than their local

counterparts. One simple example is to avoid packaged drinking water. Installation of proper water filtration facility and keeping that in working condition can reduce the tendency of buying bottled water drastically and thus can reduce ecological as well as tourist footprint of the area. Besides, stringent actions must be assured against the environmental offenders.

- Use of green fuels in kitchens should be made mandatory in hotels, resorts and homestays. Solar electricity should replace the grid supply or at least should supplement the later.
- Efficient mass transport systems should be initiated to reduce the carbon footprint of travel. Use of cleaner fuels in taxis is necessary. A better alternative for private taxies is rapid mass transport system like midi-buses that will play between all important centres and Rishyap. Per capita carbon footprint from transportation can be reduced that way.
- Establishment of a formal garbage disposal system is highly recommended at Rishyap. Hotels must implement a policy of separation of wet and dry garbage at source. On the other hand, local government must establish a garbage collection and recycling system to stop littering of hill slopes with waste materials. A complete ban on plastic bags and disposable items may be implemented.

References

Annual Final Report of Tourism Survey for the state of West Bengal (April, 2014-March, 2015) Ministry of Tourism, Govt. of India

Becken S and Patterson M, 2006: Measuring national carbon dioxide emissions from tourism as a key step towards achieving sustainable tourism. J. Sustain. Tour, 14, 323–338.

Fang K, Heijungs R, de Snoo R G Theoretical exploration for the combination of the ecological, energy, carbon, and water footprints: Overview of a footprint family. Ecol. Indic. 2014, 36, 508–518.

Galli A, Wiedmann T, Ercin E, Knoblauch D, Ewing E, Giljum S, 2012: Integrating ecological, carbon and water footprint into a footprint family of indicators: Definition and role in tracking human pressure on the planet. Ecol. Indic., 16, 100–112

Govt. of India, 2017: Annual Report, 2016-2017. Ministry of Tourism. New Delhi.

Gössling S, Hansson C B, Hörstmeier O, Saggel S, 2002: Ecological Footprint Analysis as a tool to asses tourism sustainability. Ecol, Econ., 43, 199-211.

Hadjikakou M, Chenoweth J, Miller G, 2013: Estimating the direct and indirect water use of tourism in the eastern Mediterranean. J. Environ. Manag., 114, 548–556.

Hoekstra A. Y and Champagain A. K, 2007: Water footprint of Nations: Water use by people as the function of their consumption. Water Resource Management. DOI 10.1007/s11269-006-9039-x.

Hunter C, 2002: Sustainable tourism and the touristic ecological footprint. Environ. Dev. Sustain, 4, 7-20

Joshi R and Ahmed S, 2016: Status and challenges of municipal solid waste management in India; a review. Cogent Environmental Science, Francis and Taylor Group, Vol - 2.

Manfred L et.al, 2018: The Carbon footprint of global tourism. Nature Climate Change. DOI: 10.1038/s41558-018-0141-x

Purvis, C. L. J, 2008: The ecological footprint of hostel tourists in Ontario and Quebec. International Journal of Cancer. Vol12 Issue 02, 785-790pp.

Ramachandra T.V Swetmala K and Dania T.M, 2014: Carbon footprint of the solid waste sector in Greater Bangalore, India. S.S Muthu (Ed.), Assessment of Carbon footprint in different industrial sectors. Vol - 1. Eco Productions, Springer Services and Bussiness Media, Singapore.

Rees W E, 1992: Ecological footprint and appropriated carrying capacity: What urban economics leaves out. Environ. Urban., 4, 121–130.

Sharp H, Grundius J, Heinonen J, 2016: Carbon footprint of inbound tourism to Iceland: A consumption-based life-cycle assessment including direct and indirect emissions. Sustainability, 8, 11-47

Stoeglehner G and Narodoslawsky M, 2008: Implementing ecological footprinting in decision- making processes. Land Use Policy, 25, 421-431

Sun Y, 2014: A framework to account for the tourism carbon footprint at island destinations. Tour. Manag., 45, 16–27

Wackernagel M, Rees W E, 1996: Our Ecological Footprint: Reducing Human Impact on the Earth. New Society Publishers: Gabriola Island, BC, Canada, ISBN 9780865713123

Wang S et.al, 2017: Progress and Prospects for Tourism Footprint Research. Sustainability, 9, 1847; doi:10.3390/su9101847

Yang M, Hens L, Wulf R D E, Ou X, 2011: Measuring tourist's water footprint in a mountain destination of Northwest Yunnan, China. J. Mt. Sci. 2011, 8, 682–693.

Zhang J, Zhang J, 2004: Touristic ecological footprint model and analysis of Huangshan City in 2002. ActaGeogr. Sin., 59, 763–771

Web References

http://epa.gov.uk accessed on 22.01.2019 at 7.50 pm IST.

http://data.world/footprint/nta-2018-edition accessed on 22.01.2019 at 8.30 pm IST

http://www.ibm.com/knowledge –centre accessed on 23.01.19 9.33 pm IST

http://www.wikipedia.org accessed on 24.01.2019, 9.53 pm IST.

http://www.fao.org accessed on 24.01.2019, 10.10 pm IST.

http://www.globalfootprintnetwork.org accessed on 21.07.2019 at 8.35 pm IST.

http://www.worldbank.org/en/topic/urbandevelopment/brif/solidwastemanagement accessed on 25.07.2019, 10.08 pm IST.

http://www.ibef.org/states/west-bengal.aspx#accessedon 15.07.2020, 12.00 IST.

22

Water Management and Conservation in India

Dr. Niraj Kumar
Assistant Teacher, Department of Geography, S.D. (P.G.) College,
Muzaffarnagar, Uttar Pradesh, India

Abstract

Water is a major natural resource for sustaining human existence. It only plays an important role in the cleanliness of rural and urban communities, but is also essential for all forms of agriculture and most industrial production processes but experts have always included water as one of the major resources that needs management and conservation. Those who have to be protected in future so that the population of the country can be supplied water. It would not be an exaggeration to say that water in India is on the verge of crisis rivers of the country are getting polluted; water harvesting system is deteriorating. In view of this situation the level of ground water is constantly falling, it is very important to manage water for the population in India.

Keywords: Water, Management, Conservation, Resource.

Introduction

Water is a valuable resource; it causes destruction of development somewhere in view of the increasing population and future needs the utility of every drop of water has increased. Therefore the population should be properly fed and the water resource should be used properly so that there is no shortage of water in future and people can get drop of rain on the earth so that it can be used for human work and various works, industrial works, domestic work, livestock etc. so that less in future rain famine, drought - prone, famine - prone areas can be in nature etc. for this rain water can be secured by making a dam on it, which can be in the affected areas.

Review of Studies Related to Water Management and Conservation in India

- **T.S. Chauhan (1995)** suggested that the assessment of water resources and planning and development of water resources is a fundamental component.
- **Gurjar R.K. (1990)** focused on Rajasthan's land development and studied artificial irrigation as the main basis.
- **R.P. Sen Gupta (2003)** analyzed that the role of rainwater harvesting can be played by salinity control techniques within the reservoir. It argues with mutual knowledge to manage our water resources and argues to leap. It has also been said that rainwater is stored during the monsoon.
- **Das Keshav (2006)** examines the capacity and relevance of mutual water harvesting structures in Gujarat, studying the comparative economic and social importance of mutual and modern water supply systems in arid regions.
- **Sujit Kumar C. (2002)** the reasons for the continuous decline of water level and rainwater harvesting structures in Gujarat, Rajasthan and Maharashtra. It suggested the adoption of centuries old water harvesting techniques in arid and semiarid regions.
- **Bandana Siva (2002)** attempts that water is the logical cycle ecological process through which water is obtained as an ecological system, or show. The falling moisture recharges the streams water and ground water sources. The endowment of water depends on climate, physiology, flora and geology.
- **Bhatnagar, K.P. Nigam, M.A. and Srivastava, J.P.S. (1960)** stated that the welfare of the people depends on the proper adjustment

between human i.e. an. avail as e natural resources. Natural resources like land, water, soil, minerals, human life and other living beings are balanced.

- **RB Dutta (1999)** analyzes that certain techniques and strategies for water extraction, extraction and use. He said that water resources are the most important for any form of life and it is the basis of socio-economic activities.
- **Reddy V. Ratna (2001)** studies the relationship between poverty and resource degradation and property rights and institutions in natural resource management.

Objectives

The following objectives of the study area will emerge:

1) To develop and manage water resources in India.
2) Water resources should be used as required.
3) In the context of water resources, rainwater should be stored so that it can be used as per requirement.
4) Important strategy for long term conservation and management of water resources
5) Evaluation of geological and human aspects of available water resources.
6) The demand of the present time is that the rapidly depleting water resources should be conserved and revived.
7) National Water Management Projects should be implemented by the government for water conservation and management.

Hypothesis

Water Conservation and Management Vision- It is necessary to formulate the following hypotheses for water conservation and water management, for which my hypothesis will be:

1) The growing population of India has direct relationship with water conservation and management.

2) As the population increases, so does water management and its conservation.
3) Safe use of water reserves in India and safe use of rain water. For the execution and execution of the research work presented the method, first collected data on water resources and management, which surveyed various villages and towns. For this, both primary and secondary data were analyzed and sate use of rain water.

Research Methodology

For the execution and execution of the research work presented the method, first collected data on water resources and management, which surveyed various villages and towns. For this, both primary and secondary data were analyzed. The primary data was obtained from field survey, while the second data was obtained from publication of Gratzier and statistics booklets and journals. Theoretical analysis of data was done through empirical methods. Statistics methods were also used as per requirements, which were used for water conservation and its management.

Water Management and Conservation

Water management means that water is used or disposed of properly. Water has a deep connection to life. More water is being polluted than water is used on earth and is wasted. Is being wasted. Therefore, understand the importance of water management in the future and protect it from polluting so that it can be used as needed for population fulfillment. To meet the food demand of a country with a large population like India. With water around the earth It is surrounded but only 25 percent of the water comes from natural sources, rivers, ponds and stepwells, while half percent is ground while half percent is ground water storage, 97 percent is in the sea, but it is the bitter truth that India is in the line of water crisis countries. For this water management became necessary in India, it can be managed and protected with the following measures.

Plate 1: Water Conservation
(**Source:** Google images from various Internet URLs)

Water Management Measures

The first measure of water management is to manage rain water properly. Be protected in reservoirs. Dams should be built across the country and drought the number of dams in the affected areas is high. Every drop of rainwater flowing on the land should be collected by making check dams in the rivers, to stop the water of drains. Similarly, stitches are used to replenish ground water to reach rainwater of home, to reach rainwater of home, field, village and town. Can be done by making traches and sock pits. Rainwater can be sent to places like tacos, tube wells, Bavalis etc. falling on the roofs of houses. Government efforts are necessary for water management. Different states should be motivated to conserve ground water.

Apart from this, water management can also be done with the following measures.

1. Do water harvesting at home.
2. Caution in use of water in the house.
3. Take steps to prevent uncontrolled exploitation of ground water.

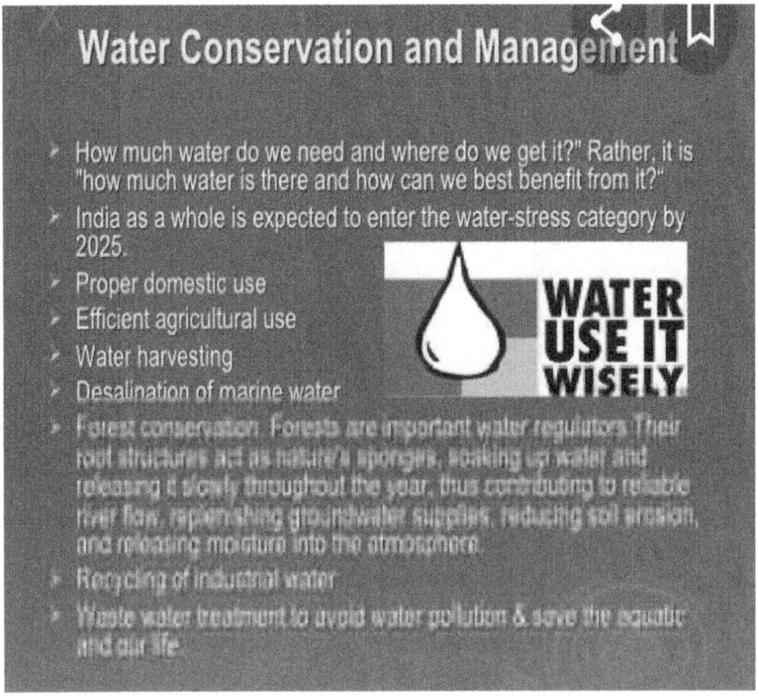

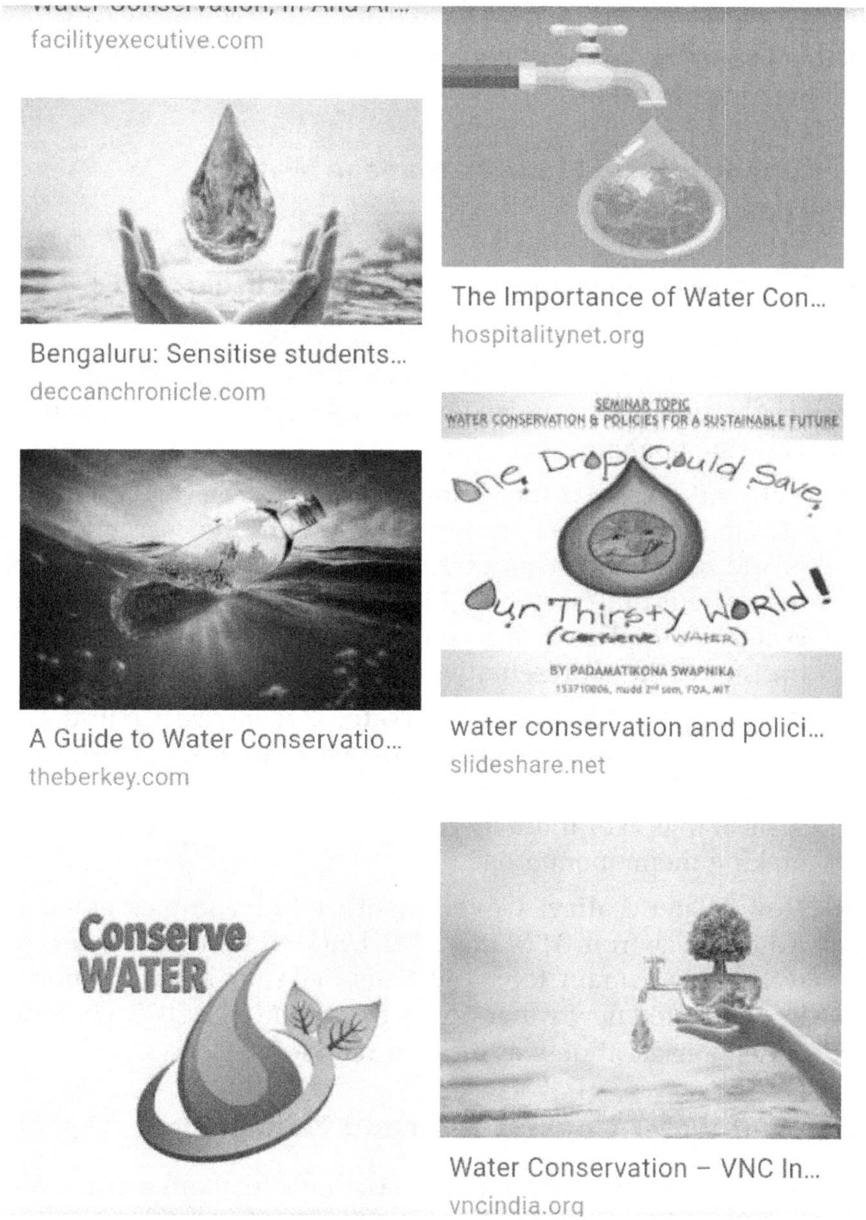

Plate 2: Water Conservation
(**Source:** Google images from various Internet URLs)

4. Indiscriminate cutting of forests should be stopped at all costs, so that the level of ground water can be stopped from decreasing it.

5. Ground water alternatives need to be explored. For this, the water available in ponds, lakes, rivers should be saved from polluting and make them mostly useful.

Government Efforts Made to Manage Water Resources in India

Water Resources Management and Training Scheme

In 1984, the Central Water Commission established the Irrigation Research and Management Organization, whose main objective was to improve irrigation development and to make the irrigation institutions efficient and sustainable.

- **National Water Management Project:** The National Water Management Project started in 1986 with the assistance of the World Bank. The major objective of this scheme was to develop the agricultural sector in India. Technology Transfer Central Water Commission has initiated a program of technology transfer to the representatives of the states.
- **Planning for Non-Physical Water Resources**: Central Water Commission has not yet done the work in those areas of India. The use of a phosphoric water is to be managed here. Like in the eastern region of India there is tube well construction work and making them operational.
- **New Water Policy**: Government of India implemented new water policy from 31 March 2002. Under this, water conservation was made a main topic. Systems of Water Conservation and implemented new water policy from 31 March 2002. Under this, water conservation was made a main topic.

Systems of Water Conservation and Management

1) In areas where the slope is not high, embankments can be installed (Mantuar Bandh), in which farmers in the rural areas will fill their fields when water over flows So, the extra water will start filling up in the field below.
2) Land dams should be constructed so that they prevent underground, drain water from entering the river channels

during the monsoon months. This can prevent the river water from getting polluted.

3) Big ponds should be constructed which is 10 m. In the areas where rainfall is less, in those areas, the left side of the catchment area should be 50 hectares in such areas.
4) Areas that are spread far and wide, special types of ponds can be made in those areas in which water leaks. These are called surface transduction rhythms.
5) Where large rivers and the annual rainfall come in them. Such areas should be modeled in those areas which have wells and ponds.
6) The nation should be made aware to the farmer that the rainwater which fills the fields. He should devise ways to prevent them from drifting into the fields.

Conclusion

The development of any region or region and country depends on the resources available in the country or territory. If there are abundant resources available in the country, then the economic development of that country will be high. In order to emerge from the water crisis in India, it is very important to manage the water so that the current population can be served. The growing population of India has an impact on water resources, given the manner in which human is misusing water, the water crisis is a very serious danger, for this water conservation is necessary in India through various measures like construction of wells, dams. Construction of water stores of rain water connecting canals and drains with ground water, etc., water management has become very necessary. Apart from this, construction of government schemes, central water storage house etc. has become very necessary. With 30 per cent of the population in India residing in cities, the report of the Ministry of Housing and Urban Development states that proper management and immediate attention is required for the waste water in about two hundred cities of the country. For this, three national water policies were made in the last 70 years of the country, the first policy was 1987, the second in 2002 and the third water policy was made in 2012. Apart from this, 14 states also formulated schemes, central water storage house etc. has become very necessary. With 30 per cent of the population in

India residing in cities, the report of the Ministry of Housing and Urban Development states that proper management and immediate attention is required for the waste water in about two hundred cities of the country. For this, three national water policies were made in the last 70 years of the country, the first policy was 1987, the second in 2002 and the third water policy was made in 2012. Apart from this, 14 states also formulated their policy which would promote water management and conservation.

References

Alam, A (2004); 'Fluoride in water'. In Survey of the Environment, The Hindu, 2004, pp. 25-28.

B.G. Verghese, (1999): Waters of Hope - from vision to reality in Himalaya - Ganga Development Cooperation, II Ed., Oxford and IBH Publishing Co. Pvt. Ltd., New Delhi, p. 96-97, 111-115.

Bansil, P.C. (2004): Water Management in India, Concept Publishing Company, New Delhi.

Bhati, Vinita, (2008): India's Water Resources Planning and Management Universal Scientific Publishers, Jaipur, Vol. 1, p. 16-32.

Bhuban C. Barar (1996): Traditional Water Harvesting System - An Ecological Economic Survey, New-age International Publishers, New Delhi, p. 1-3.

Biswas, A.K. (1976): System Approach to Water Management, Tata McGraw-Hill Publishing Company Limited, New Delhi.

Black Maggie and Talbot Rupert (2005): Water A Matter of Life and Health, Water Supply and Sanitation in Villages of India, Oxford University Press, New Delhi.

Chopra, K. et. al. (2003): Water Resources, Sustainable Livelihood and Ecosystem Services, Concept Publishing Company, New Delhi.

Das, D.C. (1977): Water harvesting for water conservation in catchment and command areas, In proceedings of the national seminar on Water Conservation and Management in Drought, IAH, New Delhi, 1-32.

Davis, K.S. and Dey, A.J. (1968): Water the Mirror of Science, Viklis, Feffer & Simphs, Bombay.

IGNP Report (2002): Water Level Study Report, submitted to Command Area Development Authority, Vol. III (WAPCOS).

Inder Jeet, (2004): Groundwater Resources of India - Occurrence Utilization and Management, Mittal Publication, New Delhi.

Inderjeet (2001): Ground Water Resources of India, Mittal Publications New Delhi. P. 1-9.

Jagadeesha, C.J. (1999) Water Resources Development and Management GIS Development, Nov.-Dec.

Jain, S.K., Kumar, Vinay (2010): Ground Water in India, A report from National Institute of Hydrology, Roorkee, pp. 40-44.

Jha, B.M. (2009): Fresh Water Status & Management, Geography and You, August, 2009, P.9-16.

Kathpalia, G.N. (1972) 'Water Policy in Agriculture', Indian Farming, Vol. XXII, No. 2, p. 17-18.

Lovovitch, M. (1977): World Water Resource, present and future, Ambio 6 (12).

Mafatlal, A.N. (1975): Social priorities for water, Commerce Annual, 3.

National Water Policy Document (2002), Ministry of Water Resources. Govt. of India.

23

Impact of Nutrients in Our Society: An Analysis

Keerthana. S.

Nutrition and Dietetics, Vellalar College for Women (Autonomous), Erode, Tamil Nadu, India

Abstract

Now a days we could not live without tasty food because we are addicted to tasty food. So, it causes many deficiency diseases. "All tasty food has no nutrients; All nutrient food has no taste". So, we should eat for alive. Not for taste. Now the most of the people can understand about the food and its health. Our India is richest country in population and it has lack of economic range. It causes major different in Indian peoples healthy and wealthy life. In this we could see about the deficiency of nutrients, how it changes the world people's life and the managements, Organizations how to help the people, future development and impact. In this world the death of the people, by deficiency of food more than the deficiency of nutrients. The only way to prevent this is to improve the quality of food, and we create the way for, it will reach to poor people and who are caused by the deficiency diseases. In many people took research about it. But no one give the treatment or solution for this. Because nature is only equal to nature. No one will exchange it.

Keywords: Deficiency of Nutrients, Statistics, People Health, Economy Impact, Organization.

Introduction

We can see the death of new born baby in every 53 seconds. What is reason? Because lack of food. Food is most essential part of our life. In ancient days Tamilans said, *"Let food to the medicine, Let medicine to the food"*

But now a days we take the medicine as our food. I think so they a practice for the future. It is about nutrients deficiency which is called malnutrition. It is in two way, that is low nutrition called malnutrition and higher nutrition also called malnutrition. And it includes the statistics of malnutrition in India. How we cured that and which cause the malnutrition. It has many reasons according to the government that is economic inequality and population. It cause the deficiency diseases like Kwashiorkor, Marasmus, gastroenteritis or chronic illness, especially the HIV/AIDS pandemic .In one side it occurs due to the lack of nutrients that is they have no proper food to give the nutrients. In another side it occurs due to Inadequate diet, overeating, lack of balanced diet. WHO said, "Malnutrition is a very big challenge to the world general health". We can use the more fertilizers to improve quantity of food but we cannot think the quality of food. Finally, here we see the solution for the malnutrition.

Malnutrition

Malnutrition or malnourishment is a condition that results from eating a diet in which nutrients are either not enough or are too much such that the diet causes health problems. One is 'under nutrition'—which includes stunting (low height for age), wasting (low weight for height), underweight (low weight for age) and micronutrient deficiencies or insufficiencies (a lack of important vitamins and minerals). Another one is overweight, obesity and diet-related non-communicable diseases (such as heart disease, stroke, diabetes and cancer).

Stunting refers to a child who is too short for his or her age. These children can suffer severe irreversible physical and cognitive damage that accompanies stunted growth. The devastating effects of stunting can last a lifetime and even affect the next generation.

Wasting refers to a child who is too thin for his or her height. Wasting is the result of recent rapid weight loss or the failure to gain weight. A child who is moderately or severely wasted has an increased risk of death, but treatment is possible.

Overweight refers to a child who is too heavy for his or her height. This form of malnutrition results from energy intakes from food and beverages that exceed children's energy requirements. Overweight increases the risk of diet-related non-communicable diseases later in life.

Definition by Gomez

In 1956, Gómez and Galvan studied factors associated with death in a group of malnourished (undernourished) children in a hospital in Mexico City, Mexico and defined categories of malnutrition: first, second, and third degree. The degrees were based on weight below a specified percentage of median weight for age. The risk of death increases with increasing degree of malnutrition. An adaptation of Gomez's original classification is still used today. While it provides a way to compare malnutrition within and between populations, the classification has been criticized for being "arbitrary" and for not considering overweight as a form of malnutrition. Also, height alone may not be the best indicator of malnutrition; children who are born prematurely may be considered short for their age even if they have good nutrition.

Table 1: Desired Body Weight According To PEM

Degree of PEM	% of Desired Body Weight for Age and Sex
Normal	90–100%
Mild: Grade I (1st degree)	75–89%
Moderate: Grade II (2nd degree)	60–74%
Severe: Grade III (3rd degree)	<60%

Source: Chowdhury, M.S.I.; Akhter, N.; Haque, M.; Aziz, R.; Nahar, N. (2009). "Serum Total Protein and Albumin Levels in Different Grades of Protein Energy Malnutrition". Journal of Bangladesh Society of Physiologist

Definition by Waterlow

John Conrad Waterlow established a new classification for malnutrition. Instead of using just weight for age measurements, the classification established by Waterlow combines weight-for-height (indicating acute episodes of malnutrition) with height-for-age to show the stunting that results from chronic malnutrition. One advantage of the Waterlow classification over the Gomez classification is that weight for height can be examined even if ages are not known.

Table 2: Desired body weight and height according to PEM

Degree of PEM	Stunting (%) Height for age	Wasting (%) Weight for height
Normal: Grade 0	>95%	>90%
Mild: Grade I	87.5–95%	80–90%
Moderate: Grade II	80–87.5%	70–80%
Severe: Grade III	<80%	<70%

Source: J.C. Waterlow (1972). "Classification and Definition of Protein-Calorie Malnutrition". British Medical Journal. 3 (5826): 566–69.

From WHO (1973)

Range of pathological condition arising from coincident lack in various proportions of protein and calories occurring mostly infants and young children and commonly associated with infections.

History

Michael Worboys, he is historian who invent the importance of malnutrition between the wars, so he invents the nutrition science. This built on work such as Casimir Funk's 1912 formulization of the concept of vitamins. There was increased scientific study of malnutrition in the 1920s and 1930s, and this became even more pronounced after world war 2. Charities and United Nations agencies would devote considerable energy to alleviating malnutrition around the world.

One of the first official global documents addressing Food security and global malnutrition was the 1948 Universal Declaration of Human Rights(UDHR). Within this document it stated that access to food was part of an adequate right to a standard of living.[143] The Right to food was asserted in the International Covenant on Economic, Social and Cultural Rights, a treaty adopted by the United Nations General Assembly on

December 16, 1966. The Right to food is a human right for people to feed themselves in dignity, be free from hunger, food insecurity, and malnutrition.[144] As of 2018, the treaty has been signed by 166 countries, by signing states agreed to take steps to the maximum of their available resources to achieve the right to adequate food.

Figure 1: It is a Sample of Awareness in Malnutrition

Types

There are two major types of malnutrition:

- Protein-energy malnutrition - resulting from deficiencies in any or all nutrients
- Micronutrient deficiency diseases-resulting from a deficiency of specific micronutrients

Protein-Energy Malnutrition

Protein malnutrition: Insufficient intake of nitrogen-containing food (protein) to maintain a nitrogen balance or nitrogen equilibrium. Children are particularly prone to develop protein malnutrition (lack of protein).It cause Kwashiorkor.

Micro Nutrient Deficiency Diseases

Micronutrient deficiencies, which is **defined** as a lack of essential vitamins and minerals required in small amounts by the body for proper growth and development. (lack of iron, vitamin -a). It causes anemia.

Signs of Malnutrition

In our body different nutrients cause different diseases. These are clearly shown in the following Table:

Table 3: Signs of Malnutrition in Various Parts of Our Body

Site	Sign
Face	Moon face (kwashiorkor), simian facies (marasmus)
Eye	Dry eyes, pale conjunctiva, Bitot's spots (vitamin A), periorbital edema
Mouth	Angular stomatitis, cheilitis, glossitis, spongy bleeding gums (vitamin C), parotid enlargement
Teeth	Enamel mottling, delayed eruption
Hair	Dull, sparse, brittle hair, hypopigmentation, flag sign (alternating bands of light and normal color), broomstick eyelashes, alopecia, and overall thinning of the hair follicles
Skin	Loose and wrinkled (marasmus), shiny and edematous (kwashiorkor), dry, follicular hyperkeratosis, patchy hyper- and hypopigmentation, erosions, poor wound healing
Nail	Koilonychia, thin and soft nail plates, fissures or ridges
Musculature	Muscles wasting, particularly in the buttocks and thighs
Skeletal	Deformities usually a result of calcium, vitamin D, or vitamin C deficiencies
Abdomen	Distended - hepatomegaly with fatty liver, ascites may be present
Cardiovascular	Bradycardia, hypotension, reduced cardiac output, small vessel vasculopathy
Neurologic	Global development delay, loss of knee and ankle reflexes, poor memory
Hematological	Pallor, petechiae, bleeding diathesis
Behavior	Lethargic, apathetic, anxious

Source: Grover, Zubin; Ee, Looi C. (2009). *"Protein Energy Malnutrition"*. Pediatric Clinics of North America.

Statistics about Nutrients Deficiency

1) There were 793 million undernourished people in the world in 2015 (13% of the total population). This is a reduction of 216 million people since 1990 when 23% were undernourished. In 2012 it was estimated that another billion people had a lack of vitamins and minerals. In 2013, protein-energy malnutrition was estimated to have resulted in 469,000 deaths—down from

510,000 deaths in 1990. Other nutritional deficiencies, which include iodine deficiency and iron deficiency anemia, result in another 84,000 deaths. In 2010, malnutrition was the cause of 1.4% of all disability adjusted life years. About a third of deaths in children are believed to be due to under nutrition, although the deaths are rarely labelled as such. In 2010, it was estimated to have contributed to about 1.5 million deaths in women and children, though some estimate the number may be greater than 3 million. An additional 165 million children were estimated to have stunted growth from malnutrition in 2013. Under nutrition is more common in developing countries. Certain groups have higher rates of under nutrition, including women—in particular while pregnant or breastfeeding—children under five years of age, and the elderly. In the elderly, under nutrition becomes more common due to physical, psychological, and social factors.

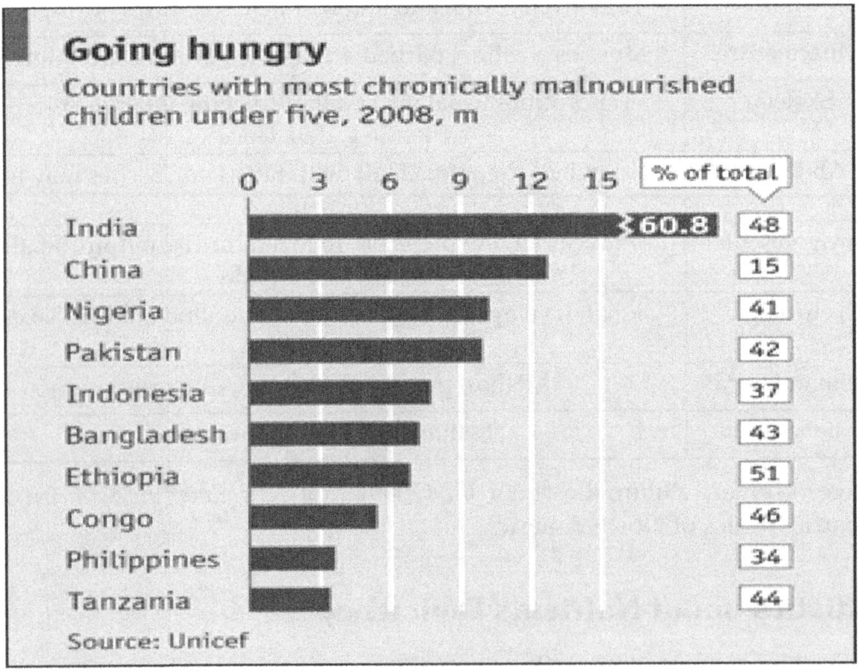

Source: Uicef statistics about child malnutrition.

Figure 2: Comparison of Malnourished Children All Over the World

2) Despite India's 50% increase in GDP since 2013, more than one third of the world›s malnourished children live in India In India

national obesity rates in 2010 were 14% for women and 18% for men with some urban areas having rates as high as 40%.

3) India is one of the fastest growing countries in terms of population and economics, sitting at a population of 1.365 billion and growing at 1.5%-1.7% annually (from 2001-2007)
4) The 2017 Global Hunger Index (GHI) Report by IFPRI ranked India 100th out of 118 countries with a serious hunger situation. Amongst South Asian nations, it ranks third behind only Afghanistan and Pakistan with a GHI score of 29.0 (serious situation).The 2019 Global Hunger Index (GHI) report ranked India 102nd out of 117 countries with a serious issue of child wasting. At least one in five children under the age of five years in India are wasted.

Table 4: Number of Undernourished Globally

Year	2005	2006	2007	2008	2009	2010	2011	2012	2013	2014	2015	2016
Number in millions	945	911	877	855	840	821	813	806	795	784	784	804
Percentage	14.5%	13.8%	13.1%	12.6%	12.2%	11.8%	11.5%	11.3%	11.0%	10.7%	10.6%	10.8%

Source: "The State of Food Insecurity in the World 2018" (technical annex) From, Food and Agricultural Organization of the United Nations. Retrieved January 11, 2019

5) 821 million undernourished people in the world in 2018 (10.8% of the total population). This is a reduction of about 176 million people since 1990 when 23% were undernourished, but an increase of about 36 million since 2015, when 10.6% were undernourished In 2012, it was estimated that another billion people had a lack of vitamins, and minerals. In 2015, protein-energy malnutrition was estimated to have resulted in 323,000 deaths—down from 510,000 deaths in 1990. Other nutritional deficiencies, which include iodine deficiency and iron deficiency

anemia, result in another 83,000 deaths. In 2010, malnutrition was the cause of 1.4% of all disability adjusted life years. About a third of deaths in children are believed to be due to under nutrition, although the deaths are rarely labelled as such In 2010, it was estimated to have contributed to about 1.5 million deaths in women and children, though some estimate the number may be greater than 3 million. An additional 165 million children were estimated to have stunted growth from malnutrition in 2013.

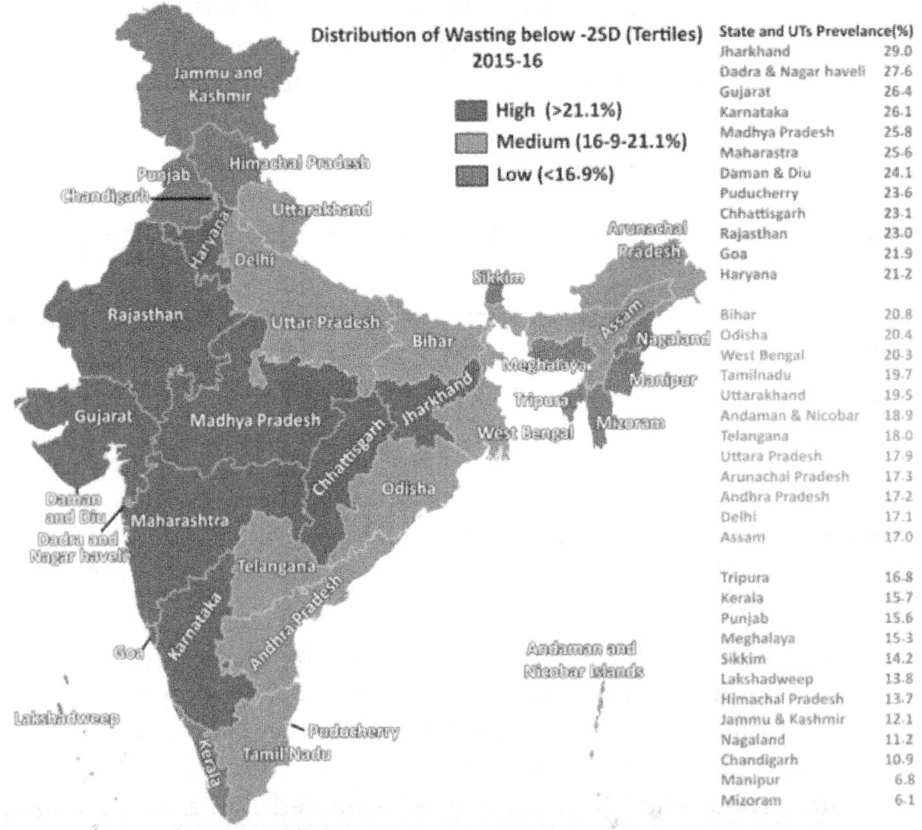

Source: A Screenshot of the atlas's dashboard displaying the distribution of wasting in India's status.

Map 1: Distribution of Wasting

Table 5: Number of Undernourished in The Developing World

Year	1969-71	1979-81	1990-92	1995-97	2001-03	2003-05	2005-07	2010-12
Number in millions			820	790	825	848	927	805
Percentage	37%	28%	20%	18%	17%	16%	17%	14%

Source: "The State of Food Insecurity in the World 2006"(technical annex). From, Food and Agricultural Organization of the United Nations. Retrieved April 6, 2019

6) In India 44% of children under the age of 5 are underweight. 72% of infants and 52% of married women have anemia. Research has conclusively shown that malnutrition during pregnancy causes the child to have increased risk of future diseases, physical retardation, and reduced cognitive abilities.

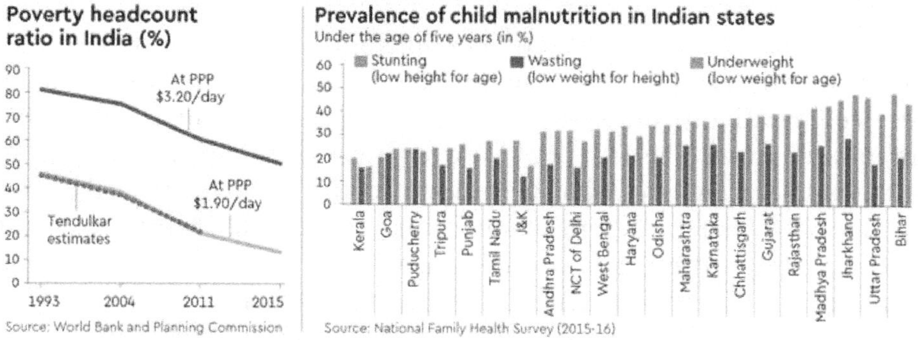

Source: World bank and **Source:** National family health survey (2015-2016) planning commission.

Figure 3: Child Malnutrition in Indian States

Theories about Malnutrition in Various Views

India is often thought of as a development paradox with relatively high economic growth rates in the past few years, but with lower progress in areas of life expectancy, education and standard of living. While serious inequalities in growth, development and opportunity explain the illusion of the paradox at the country level, still, a significant proportion of the world's poor live in India, as do a significant proportion of the world's malnourished children. Poverty and under nutrition coexist, and poor dietary quality is associated with poor childhood growth, as well as significant micronutrient deficiencies. Food security is particularly vulnerable to changes in the economic scenario and to inequities in wealth distribution. Migration from rural to urban settings with a large informal employment sector also ensures that migrants continue to live in food insecure situations. While food production has for the most part kept pace with the increasing population, it has been with regard to cereal rather than of pulses and millet production. Oil seeds, sugar cane and horticultural crops, along with non-food crops are also being promoted, which do not address nutrition security, and, coupled with the increase in the consumption of prepared food, may indeed predispose towards the double burden of malnutrition. Access to food is also particularly susceptible to poverty and inequality. Many strategies and policies have been proposed to counter under nutrition in India, but their implementation has not been uniform, and it is still too early to assess their lasting impact at scale.

The relationship between malnutrition, food insecurity, and poverty in older persons from urban Latin America has, to date, received relatively little attention. This thesis aims to address this important issue in the setting of Mexico City and its Metropolitan Zone. A theoretical framework has been developed to understand the causal linkages between the determinants of malnutrition, food insecurity and poverty, and current data are then used to describe how this public health concern manifests itself in urban Latin American contexts. This is followed by a discussion of recent social policy interventions aimed at improving nutrition, access to food and well-being in older people in Latin America. The quantitative part of the thesis presents a study which assessed indicators of nutritional status, food security, health, quality of life and living conditions among 1,263 households with residents aged 70 and over from socio geographically- defined poor areas of Mexico City and its Metropolitan Zone. The literature suggests that a regular source of economic resources is important to ensure food security during old age; thus, the impact of an ongoing old-age monetary-transfer program on nutrition-related indicators, food security and poverty was assessed at the levels of both older persons and households. This was carried out through a quasi-experimental study using an ex-post comparison of intervention and control groups with no baseline measures. Overall, results suggest differential access to food, quality of life and living conditions according to socioeconomic stratum among older persons and their household contexts. Differences in dietary diversity and food insecurity among older persons were also found, when data were disaggregated by monetary transfer eligibility status. The old-age intervention analyzed in this thesis showed little impact on access to food and other indicators of well-being at household level. There was, however, a high prevalence of people being overweight and obese among the older population under study. Given that this thesis is the first approach to food insecurity ever carried out among urban older populations in Mexico and Latin America, the conclusions emphasized the magnitude of uncertain access to food during old age, and they suggest guidelines for policy makers at different levels of government, stimulating further research on issues related to old age in the region.

Case Study

The following case study from Tamil Nadu, a southern state of India focuses on the complex challenges faced and the progress made so far as part of efforts towards combating malnutrition. It also demonstrates how lessons are being learned along the way.

1. The Tamil Nadu integrated nutrition project (TINP)

- The Tamil Nadu Integrated Nutrition Project (TINP), a World Bank assisted intervention program in rural south India, offered nutrition and health services to children under-5 and pregnant and lactating women. TINP-I (1980–1989) eventually covered 174 blocks. It was a forerunner of the Bangladesh Integrated Nutrition Project (BINP). TINP-II (1991–1997) covered all non-ICDS blocks in the Tamil Nadu state. TINP-II was replaced by World Bank assisted ICDS III (WB-ICDS III) from 1998.
- Since 1975, Indian government is providing a package of services to combat child hunger and malnutrition under Integrated Child Development Services (ICDS) program through Anganwadi centers (AWCs). Anganwadi means 'courtyard shelter' in local language.

2. TINP I (1980–1989)

- Approximately 1.25–2.40% points per year (ppt/year) drop in underweight prevalence was noted among beneficiaries. On comparing drop in underweight prevalence between TINP areas and non-TINP areas, it was noticed that drop was approximately 0.83–1.12 ppt/year in TINP areas whereas reduction in underweight prevalence was approximately 0.26–1.12 ppt/year in non-TINP areas.

- At the same duration, reduction in the underweight prevalence was estimated as 0.7 ppt/year for the whole of India. Therefore, it can be stated that quarter to half of the reduction in underweight prevalence was attributable to the TINP project.
- Having achieved a significant reduction in severe early childhood malnutrition, TINP-1 became inspiration for others as a 'success story' during the 1980s. Evaluations indicated a decrease in underweight prevalence of about 1.5% points per year in participating districts, twice the rate in non-participating districts. Several factors contributed in the success story of TINP I viz. selective feeding (the careful focus on supplementing the dietary intake of young children when their growth faltered and until their growth resumed), clarity in job responsibilities and description, positive worker-supervisor ratio and robust monitoring system.

3. TINP II (1990–1997)

- TINP II was rolled out to move beyond reducing severe malnutrition and with a more ambitious objective to significantly reduce the burden of moderate malnutrition. In other words, it shifted towards a more preventive focus. Core strategies adopted in TINP II were regular growth monitoring, nutrition education, health check-ups, supplementary feeding of malnourished children and growth-faltering children, high-risk pregnant and lactating women.
- Approximately 6.0 ppt/year drop in underweight prevalence was noted among TINP II beneficiaries. It was also noticed that drop was approximately 1.1 ppt/year in TINP areas. As per estimates of World Bank, the current underlying trend in the state was to be 5.0–7.0 ppt/year, which is most certainly an overestimate.

4. Missions for Prevent the Disorders

- The Food and Agriculture Organization of the UN has created a partnership that will act through the African Union's CAADP framework aiming to end hunger in Africa by 2025. It includes different interventions including support for improved food production, a strengthening of social protection and integration of the right to food into national legislation
- The Ending Hunger campaign is an online communication campaign aimed at raising awareness of the hunger problem. It has many worked through viral videos depicting celebrities voicing their anger about the large number of hungry people in the world.
- After the Millennium Development Goals expired in 2015, the main global policy focus to reduce hunger and poverty became the Sustainable Development Goals. In particular Goal 2: Zero hunger sets globally agreed targets to end hunger, all forms of malnutrition and promote sustainable agriculture. The partnership Compact2025, led by IFPRI with the involvement of UN organizations, NGOs and private foundations develops and disseminates evidence-based advice to politicians and other decision-makers aimed at ending hunger and under nutrition in the coming 10 years, by 2025
- Another way is controlled population, Thomas Malthus argued that population growth could be controlled by natural disasters and voluntary limits through "moral restraint."

Managements and Organizations to Help the People

Midday Meal Scheme in Indian Schools

The Indian government started midday meal scheme on 15 August 1995. It serves millions of children with fresh cooked meals in almost all the government run schools or schools aided by the government fund.

Apart from this Food for Life Annamrita run by ISKCON Food Relief Foundation and the Akshaya Patra Foundation run the world›s largest NGO-run midday meal programmes, each serving freshly cooked plant-based meals to over 1.3 million school children in government and government-aided schools in India.

Integrated Child Development Scheme

- communication for behaviour change
- capacity building
- community's access to tangibles and intangibles
- community participation and
- collective approach.

National Children's Fund

The National Children's Fund was created during the International Year of the Child in 1979 under the Charitable Endowment Fund Act, 1890. This Fund provides support to the voluntary organizations that help the welfare of kids

National Plan of Action for Children

India is a signatory to the 27 survival and development goals laid down by the World Summit on children 1990. In order to implement these goals, the Department of Women & Child Development has formulated a National Plan of Action on Children. 15 State Govts have prepared State Plan of Action on the lines of National Plan of Action specifying targets for 1995 as well as for 2000 and spelling out strategies for holistic child development

United Nations Children's Fund

Department of Women and Child Development is the nodal department for UNICEF. India is associated with UNICEF since 1949 and is now in the fifth decade of cooperation for assisting most disadvantaged children and their mothers.

National Rural Health Mission

The National Rural Health Mission of India mission was created for the years 2005–2012, and its goal is to «improve the availability of and access to quality health care by people, especially for those residing in rural areas, the poor, women, and children.»

The subset of goals under this mission are:

- Reduce infant mortality rate (IMR) and maternal mortality ratio (MMR),Neonatal mortality rate(NMR).

- Provide universal access to public health services
- Prevent and control both communicable and non-communicable diseases, including locally endemic diseases
- Provide access to integrated comprehensive primary healthcare
- Create population stabilization, as well as gender and demographic balance
- Revitalize local health traditions and mainstream AYUSH
- Finally, to promote healthy life styles

Establishment of Nutritional Rehabilitation Centers (NRCs)

NRCs have been set up at facility level to provide medical and nutritional care to Severe Acute Malnourished (SAM) children under 5 years of age who have medical complications. In addition, the mothers are also imparted skills on child care and feeding practices so that the child continues to receive adequate care at home

Some other programs:

- Promotion of Infant and Young Child Feeding practices (IYCF)
- Anemia Mukt Bharat (AMB)
- National De-worming Day (NDD)
- Village Health and Nutrition Days (VHNDs)
- Rajiv Gandhi National Creche Scheme
- Sarva Shiksha Abhiyan (SSA)

Conclusion

Through this we could see about the problem of malnutrition. The most problem is economic inequality. If we manage the economies and population in India, we definitely control the death and loss of people through the deficiency of nutrition. And increase the quality of nutrients in foods through avoid the artificial fertilizers. In upcoming days, we must face the problem according our health, so we have to be prepared to fight against and improve our boosting system, for nature disorders. From any disorder not only health issues and also like food, water problem we must save the nature and It is our duty. *If we save the nature, the nature will save us.*

Reference

Ackerson, L. K.; Subramanian, S. V. (15 May 2008). *"Domestic Violence and Chronic Malnutrition among Women and Children in India"*. American Journal of Epidemiology. 167 (10): 1188–1196. doi:10.1093/aje/kwn049.

Balarajan, Yarlini; Reich, Michael R. (1 July 2016). "Political economy of child nutrition policy: A qualitative study of India's Integrated Child Development Services (ICDS) scheme". *Food Policy*. 62: 88-98. doi:10.1016/j.foodpol.2016.05.001.

J.C. Waterlow (1972). "Classification and Definition of Protein-Calorie Malnutrition". *British Medical Journal*. 3 (5826): 566–69. doi:10.1136/bmj.3.5826.566.

Kanjilal, Barun; Mazumdar; Mukherjee; Rahman (January 2010). "Nutritional status of children in India: household socio-economic condition as the contextual determinant". *International Journal for Equity in Health*. 9:19-31. doi:10.1186/1475-9276-9-19.

Rivera-Marquez, Jose Alberto; (2006) Malnutrition, food insecurity and poverty in older persons from Mexico City. PhD thesis, London School of Hygiene & Tropical Medicine. DOI: https://doi.org/10.17037/PUBS.01343271.

Thow, Anne Marie; Kadiyala, Suneetha; Khandelwal, Shweta; Menon, Purnima; Downs, Shauna; Reddy, K. Srinath (June 2016). "Toward Food Policy for the Dual Burden of Malnutrition: An Exploratory Policy Space Analysis in India" .*Food and Nutrition Bulletin*. 37 (3): 261-274. doi:10.1177/0379572116653863.

UNICEF. *"Introduction to Nutrition"* from, *"Children, food and nutrition"* (34-48,94-112).

Varadharajan, Kiruba S; Thomas, Tinku and Kurpad, Anura V. Poverty and the state of nutrition in India [online]. *Asia Pacific Journal of Clinical Nutrition*, Vol. 22, No. 3, 2013: 326-339.

Wellman, N.S; Weddle, D.O, Kranz, S, Brain, C.T (October 1997). "Elder Insecurities: Poverty, Hunger, and Malnutrition". *Journal of the American Dietetic Association* 97 (10): S120–S122. doi:10.1016/S0002-8223(97)00744-X.

24

An Assessment of the Status of Household Solid Waste and its Management: A Case Study of Jalpaiguri Municipality, West Bengal, India: An Insight

Debarati Chowdhury[1] & Dr. Indrajit Roy Chowdhury[2]

[1]Ph.D. Research Scholar & [2]Assistant Professor,
Department of Geography & Applied Geography,
University of North Bengal, Siliguri, West Bengal, India

Abstract

The disposable solid materials are known as solid waste and human beings are the greatest generator of such waste. Uncontrolled growth of population in recent years and ramifications of urbanisation has made solid waste management an important issue. Due to lack of serious efforts by town/city authorities, the management of solid waste has become a tenacious problem now-a-day. Most of the towns and cities of west Bengal has labour intensive technology of waste disposal without any treatment and recovery. This study intended to provide an assessment of household waste management of Jalpaiguri, a census town of West Bengal. There are variety of sources of waste generation in this town, huge concentration of population of different income groups, generates a large quantity of waste every day in their own ways. A large space is occupied by household sector

[1]Corresponding Author

in this town which contribute highest share of daily waste generation. The study is mainly based on primary data which is collected from 50 households of the municipality and is aimed at assessing current status on household waste collection, transportation, disposal, questions of perception, and the means of engagement of solid waste management at household level. It also dealt with some identified problems and based on this some recommendations were suggested for better management. Questionnaire was framed for collecting the primary data. Data obtained from questionnaire has been computed and analysed with the help of statistical application, cartographic representation. F test has been carried out for showing the relation between household family size and waste generation. Factor analysis has been used for assessment of predominant factor between environmental impact of waste and solid waste management strategy.

Keywords: Household Waste, Waste Segregation, Landfill, Reduce, Reuse, Recycle.

Introduction

Population increase and limited resources in developing countries are curtailing solid waste management systems with improper planning (*Geeta Mohan Kulshrestha, 2007*). At the early stages of human civilization, disposal of solid waste was not much of a problem as there were a lot of open spaces and very low population.

But gradually as the population increases at faster rate and the amount of open spaces are decreases, the disposal of solid waste becomes a genuine matter of concern. Several studies indicate that much of municipal solid waste generates from households (55-80%), market areas (10-30%) and other varying quantities from industries, institutions etc. (*Kodwo Miyeza et al., 2015*). The annual waste generation is continuously rising as a result of which solid waste are disposed in unscientific manner often poses environmental threat. (*Benneworth, 2009*). India is currently confronting several challenges regarding solid waste management dilemma. Even though there is a legislative framework existing to address municipal solid waste management, open dumping is common problem and there is no segregation of waste. Besides, community sensitization is low. (*Kumar and Tayade, 2008*). Jalpaiguri, a town of West Bengal, is not an exception in this context. In Jalpaiguri town, approximately 12552 MT waste generates per day. (*Arindam Ghosh et al., 2019*) Jalpaiguri Municipal Corporation is responsible for collection, transportation and disposal of solid waste. This article is mainly concerned with domestic waste; the household has taken as a sample unit. Therefore, the study addressed on household waste management of Jalpaiguri municipal area i.e. the quantity and nature of waste generation and their treatments. Household's waste disposal practises are analysed and presented. The study also demonstrates households' attitudes and behaviour with regard to waste management.

Legislation of Solid Waste Management in India

The Union Ministry of Environment Forests and Climate Change (MoEF & CC) notified new Solid Waste Management Rules in 2016. These replaced the Municipal Solid Wastes (Management and Handling) Rules, 2000. These new rules are applicable beyond municipal area and included all the census towns, urban agglomerations, notified industrial townships and so on. As per these rules-

Duties of Waste Generators: Every waste producer should-

- Accumulate their waste in three separate wastes namely bio-degradable, Non-biodegradable; domestic hazard wastes in bin and handover the segregated wastes to the waste collectors.
- No waste generator should throw burn or bury solid wastes on streets and open public spaces.
- Storing of horticulture waste and garden waste and disposing of as per the direction of local body from time to time.

Responsibilities of Local Authorities and Urban Agglomerations: the local authorities should-

- Arrange door to door collection of waste from all households including informal settlements.
- Not to dispose of any waste on open spaces, drains, water bodies and to segregate the waste at source.
- Set up material recovery facilities for sorting of recyclable materials.
- Provide training on solid waste management to rag pickers and waste collectors.
- Ensure safe storage and transportation of the domestic hazardous waste to hazardous waste.

- Promote setting up of decentralized compost plan and or bio-methanation plan at suitable location etc.

The implementation schedule of these Municipal Solid Waste Management Rules, 2016 has been shown in following table: -

Table 1: Implementation Schedule of Municipal Solid Waste Management Rules, 2016

Activity	**Time limit from the date of notification of rules**
Identification of suitable sites for setting up solid waste processing facilities	1 year
Identification of suitable sites for setting up sanitary landfill having a population of 0.5 million or more	1 year
Conduction of door to door collection of segregated waste and its transportation in covered vehicles	2 years
Enforcing waste producers to practice segregation of different waste at source	2 years
Setting up solid waste processing facilities by all local bodies and census towns below 100000 population	3 years
Setting up solid waste processing facilities by all local bodies having 100000 or more population	2 years

Source: CPCB, 2016

Overall Scenario of Solid Waste Management of Different Census Towns in West Bengal

Solid Waste management is a major obligatory function of urban local bodies. (*Rene Veron and Anna Zimer, 2006*). When we examine the solid waste management practices in west Bengal it is observed that the exact quantity of waste generated in west Bengal is not totally known except for certain micro level studies conducted in different census towns and municipalities. (*K. Naresh Kumar and Sudha Goel, 2008*). Continuous urbanisation and consumption of goods make it a tedious task and owing to this, the composition of waste also varies. Among all the municipal corporation Kolkata MC, Asansol MC, Howrah MC having more than 10 lakh population produces average 3000-4000 MT waste per day, whereas urban local bodies of Kolkata Metropolitan Area produce 150-200MT waste per day. (*Kurian Joseph, 2002*). Door to door waste collection system is available across 60% of the census towns in West Bengal. Wheelbarrows, tricycles and pushcarts are the major waste collection vehicles. Segregation of waste are not performed at the time of collection. The number of people involved in door to door waste collection varied from as low as 13% of required numbers to 90% of required capacity (*Urban Development & Municipal Affairs Department, Govt. of West Bengal, 2011*). Street sweeping at least once a day is carried out in most of the municipalities. There is a lack of recycling of waste and very limited portion of recyclable matter actually ends up at recycling. The two bin systems which are being increasingly adopted in many countries in recent years, have largely not taken up in West Bengal. Some of the municipalities like Diamond harbour, Maheshtala, Baruipur, Konnagar, Dumdum, Howrah (Bally) set up material recover facility and vermicomposting plan, most of them are non-functional. (*M. Sanyal et al., 2010*).

Study Area

Jalpaiguri municipal area is selected for the present study. Jalpaiguri is located at 26°31′12″N and 88°43′48″E. It lies 35 km away from Siliguri metropolis. It is situated on the banks of Tista River, lying in the middle of Terai plains. Tista and Karala River bisect the town into two halves. According to 2011 census, Jalpaiguri had a population of 169,002.

Significance of the Study

Solid waste management is becoming very complicated due to urbanisation and populating increase in recent years. It creates several problems in waste management as the town grows in an unplanned and haphazard way. (*Aparesh Mondal, 2018*). Jalpaiguri is such a town that is growing in all aspects as it is well connected with roadways and railways. Since the advent of twenty century this town has been witnessing rapid development in segments like education, information technology firm, medical facilities and administrative facilities etc. (*Mahasweta Chakraborty, 2020*).

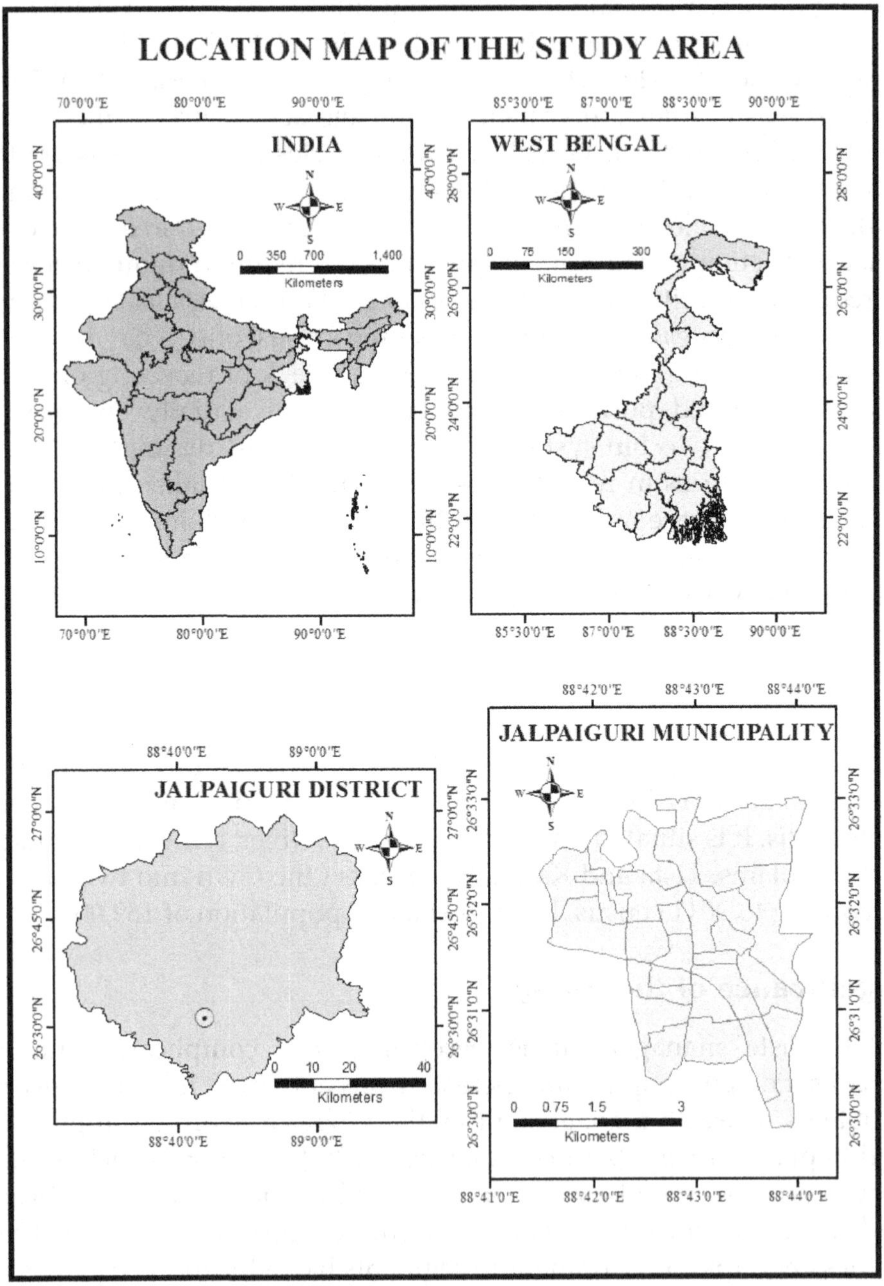

Source: NATMO and Jalpaiguri Municipality

Figure 1: Location Map of the Study Area

It is an important town of Terai region of West Bengal and also represents as one of the eco-friendly towns. But solid waste is an emerging problem of this town and it suffers from lots of waste generation with improper management which threatens the urban environment. (*Arundam Ghosh et al., 2019*) Therefore, an endeavour has been made in this study to discuss the methods, practices associated with sources, disposal of solid waste made by the households of Jalpaiguri town based on different socio-economic variable like occupation, family size, type etc. Household are taken as a sample unit as it contributes largest share of daily waste generation.

Solid Waste Management Practices in Jalpaiguri Municipality

Jalpaiguri municipality currently consists of 25 wards and 169, 002 population (Census of India, 2011). The numbers of registered households are occupied most of the municipal area. This makes it difficult to the municipal workers to implement solid waste management across all parts of Jalpaiguri town.

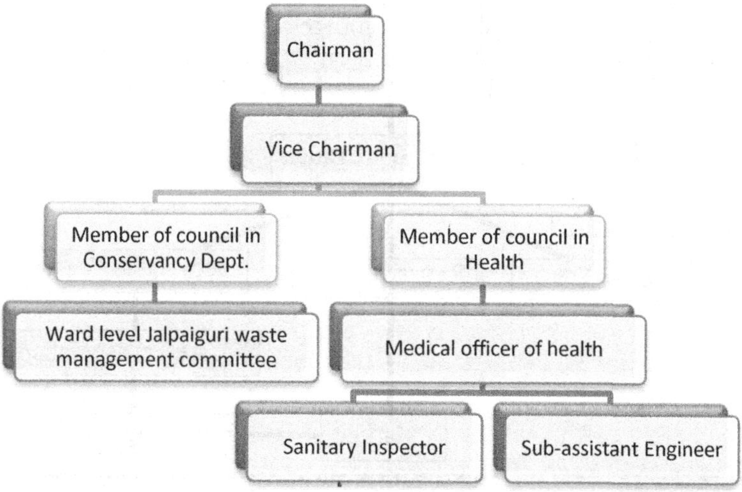

Source: Jalpaiguri Municipality

Figure 2: Institutional Framework of Solid Waste Management in Jalpaiguri Municipality

Institutional Framework of Solid Waste Management: At municipal level of Jalpaiguri, the planning, implementation of management of solid waste is based on a hierarchical system. Decisions on solid waste management are undertaken by chairman in council and circulate to sanitary inspectors and sub assistance engineers. Under them, team of conservancy inspectors, sweepers and daily wage workers are appointed. The institutional framework found in Jalpaiguri Municipality is shown below.

Waste Collection Services: In Jalpaiguri, municipal workers are appointed under each ward by the municipal authority for collecting household waste. Therefore, household members hand over the waste whenever they come at their houses for collecting waste. There are 30 community bins present in Jalpaiguri. (Jalpaiguri municipality, 2011) Wastes from markets, hotels, restaurants, commercial areas are accumulated in community bin.

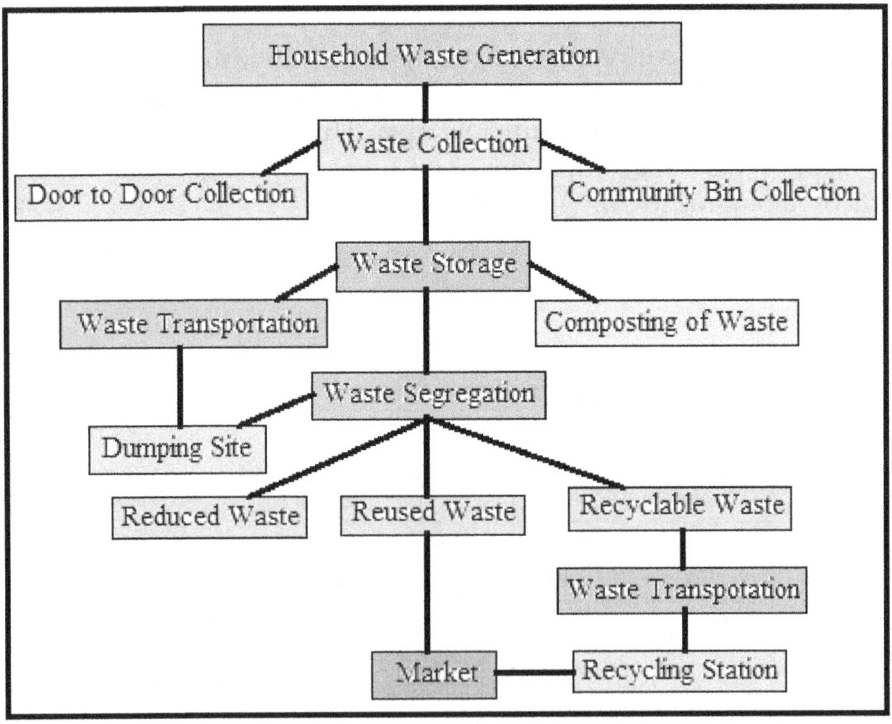

Source: Jalpaiguri Municipality

Figure 3: Schematic Diagram of Household Solid Waste Management

Waste Transportation: The municipal authorities fail to adequately provide specially designed hi tech vehicles for waste transport in Jalpaiguri. Household wastes are carried out by 4 numbers of ordinary trucks and 3 numbers of tractors. 30 tricycle vans and 10-wheel burrows are purchased for collecting waste.

Table 2: Availability of Waste Transportation in Jalpaiguri Municipality

Sl. No.	Particulars	Nos. Available
1	Hydraulic Dumper	NIL
2	Hydraulic Dumper Placer	NIL
3	Covered Truck	NIL
4	Dumper Truck	NIL
5	Ordinary Truck	4
6	Refuse Trailer	NIL
7	Tractor	3
8	Try-cycle Van	30
9	Wheel Barrow	10
10	Bulldozer / Pay-loader	NIL
11	Community Bins	30
12	Hydraulic Dumper Bins	NIL
13	Cess Pool Emptier	1

Source: Jalpaiguri Municipality.

Disposal of Waste: There is no proper dumping ground in the town. Wastes are usually dumped at *Panga* dumping ground, which is located outside of the main town at a distance of 12 km from the centre. One dumping site is developed on Balapara, operating since 2013, covering 2.19-acre area, and about 48 MT tons per day wastes are dumped there every day.

Objectives of the Study

The purpose of the study is to fulfil following objectives-

- To assess the quantity and types of solid waste produced at household level of Jalpaiguri town.
- To study the existing status of household waste collection, storage, treatment and disposal activities.

- Identifying the problems facing by the households due to untreated waste of the study area.
- To measure the awareness and perception of the households about the better management practices of the study area.
- To suggest some recommendations for better management options and environmental problems in Jalpaiguri municipal area.

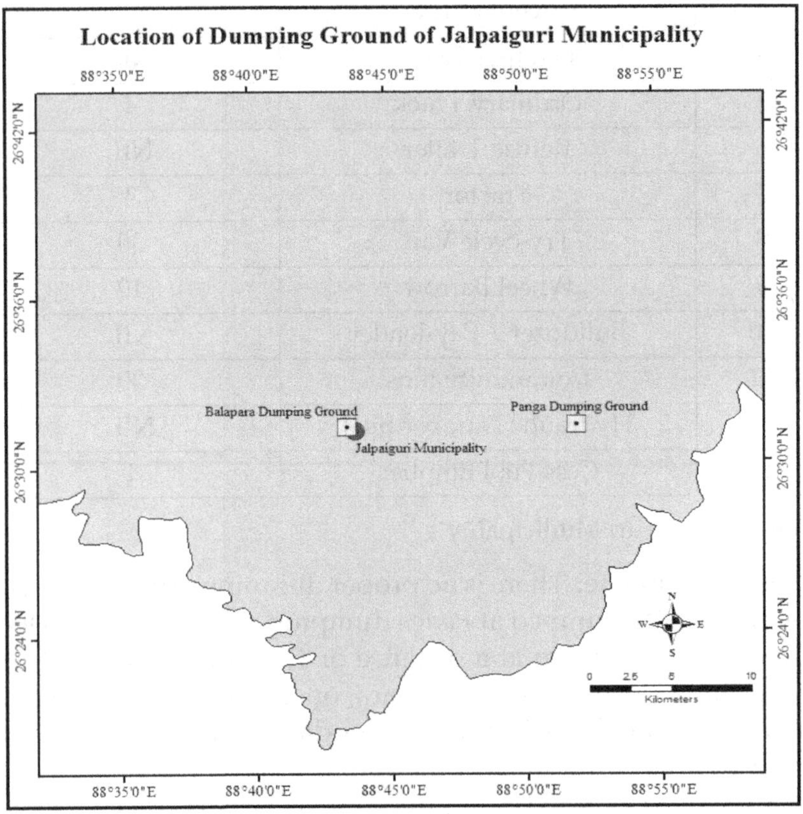

Source: Collected from Google Earth & represented by the authors

Figure 4: Location of Dumping Site of Jalpaiguri Municipality

Methodology Adopted for the Study

The entire methodology has been divided into 3 phases for the study:

- **First Phase:** In the first phase of the study, an attempt has been made to collect several information from the secondary data sources related to solid waste management. Information from national and international journals, thesis papers have been thoroughly consulted in this regard. A structured questionnaire was framed for collecting the primary data regarding household profile, waste generation and types, waste collection and disposal practices, households' attitude towards waste management etc.

- **Second Phase:** Sample survey has been carried out for collecting the primary data in the second phase. Due to unavailability of secondary data from Jalpaiguri municipality office, the study is mainly relied upon primary data. Simple random sampling method was opted in course of survey. 50 households were randomly chosen for conducting the survey.
- **Third Phase:** At the final stage data obtained from questionnaire has been computed and analysed with the help of statistical application, cartographic representation in order to illustrate present status of household waste management scenario of the study area.

Result and Discussion

Household Profile

To assess the status of household waste management in Jalpaiguri municipal area it is important to know respondent's education, occupation, family size, income etc, these parameters help us to understand their cognition level regarding waste management.

Table 3: Educational Qualification

Primary Schooling	Secondary Schooling	College or Higher
10.26%	15.38%	74.36%

Source: Primary Data, 2020

Education plays a crucial role in decision making process because with the increase level of education awareness towards waste management increases. It is evident from the table 1.3 that around 74.36% of respondents are graduates and well educated, while the others of which, 10.26% are primary school passed and 15.38% completed high school in the study area.

Table 4: Occupational Status

Govt. Employee	Private Employee	Business	Retired	Others
33.33%	20.51%	30.77%	2.56%	12.82%

Source: Primary Data, 2020

The respondents irrespective their occupations influence waste generation in the study area. 33.33% of respondents are engaged in govt. job, 20.51% involved in private jobs, 30.77% are associated with business and 2.56% of respondents are retired.

Table 5: Family Size

<3	3	4	5	6	>6
5.13%	30.77%	35.90%	12.82%	7.69%	7.69%

Source: Primary Data, 2020

Size of the family is most appropriate factor in the estimation of household waste. It is found from table 1.5 that majority of the respondent households belong to medium sized families with four members (35.90%) and three members (30.77%). The large families with six and more than six members accounted for only 7.69% of the total respondent households.

Table 6: Monthly Household Income

<10000	10000-20000	20000-30000	30000-40000	>40000
10.26%	25.64%	15.38%	23.08%	25.64%

Source: Primary Data, 2020

Income is the important factor for understanding the consumption pattern of the households as it decides the purchasing power of the households. It is seen from the table 1.6 that high, medium and low all three income groups are present in the study area.

Household Waste Generation and Disposal

On the basis of the information furnished by 50 households of Jalpaiguri municipal area in the given questionnaire quantity of waste generation, waste disposal methods, waste types etc. have been discussed and shown in the following figures-

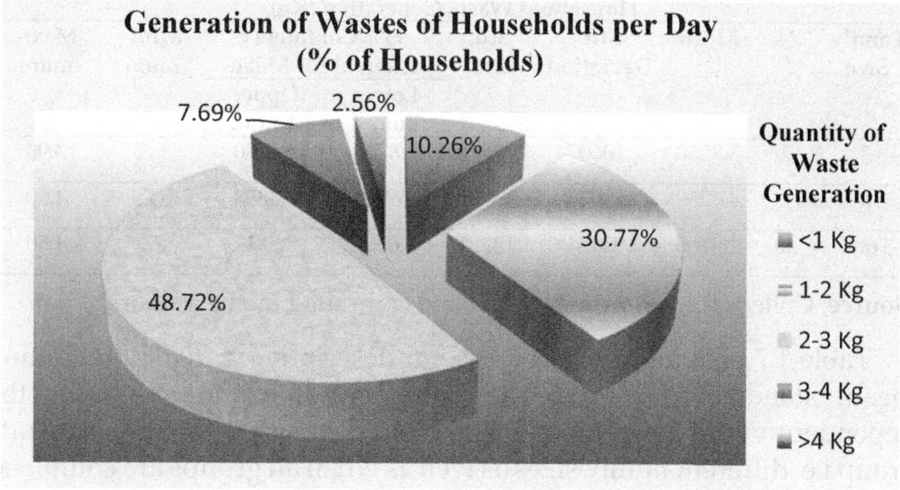

Source: Primary Data, 2020

Figure 5

A look at the data pertaining to household waste generation of Waste as reported by respondents (Fig 1.5) shows that 48.72% households generates 2-3 kg wastes per day, 30.77% of households produces 1-2 kg wastes per day and only 10.26% households produces less than 1 kg waste per day. There are negligible percentage of households produce 3-4 and more than 4 kg wastes per day.

Table 7: Statistical Analysis of Relation between Family Size & Generation of Waste- Descriptive

Household Waste Generation (Kg)								
Family Size	N	Mean	Std. Deviation	Std. Error	95% Confidence Interval for Mean		Mini-mum	Maxi-mum
					Lower Bound	Upper Bound		
2	4	1.0125	.43277	.21639	.3239	1.7011	.60	1.50
3	14	1.4429	.34965	.09345	1.2410	1.6447	.80	2.25
4	17	2.4706	.51903	.12588	2.2037	2.7374	1.50	3.50
5	7	2.9286	.05669	.02143	2.8761	2.9810	2.85	3.00
6	4	3.2875	.47324	.23662	2.5345	4.0405	2.90	3.95

Household Waste Generation (Kg)								
Family Size	N	Mean	Std. Deviation	Std. Error	95% Confidence Interval for Mean		Minimum	Maximum
					Lower Bound	Upper Bound		
7	2	3.8250	.10607	.07500	2.8720	4.7780	3.75	3.90
8	2	4.1250	.53033	.37500	-.6398	8.8898	3.75	4.50
Total	50	2.3160	.93484	.13221	2.0503	2.5817	.60	4.50

Source: Collected from primary data and computed by the authors

Table 1.7 provide some very useful descriptive statistics, including mean, standard deviation and 95% confidence intervals for the dependent variable i.e. household waste generation for each separate group i.e. different family sizes as well as when all groups are combined (total).

Table 8: Statistical Analysis of Relation between Family Size & Generation of Waste- ANOVA (F Test)

Household Waste Generation (Kg)					
	Sum of Squares	df	Mean Square	F	Sig.
Between Groups	35.377	6	5.896	34.054	.000
Within Groups	7.445	43	.173		
Total	42.822	49			

Source: Collected from primary data and computed by the authors

The above table (Table 1.8) shows the output of the ANOVA analysis and whether there is a statistically significant difference between the group means. It can be seen that the significance value is .000, which is less than the F value i.e. 34.054 and therefore, there is a statistically significant difference in the mean household waste generation between the different family sizes. Thus, it can be clearly stated that amount of household waste generation depends on family sizes in the study area.

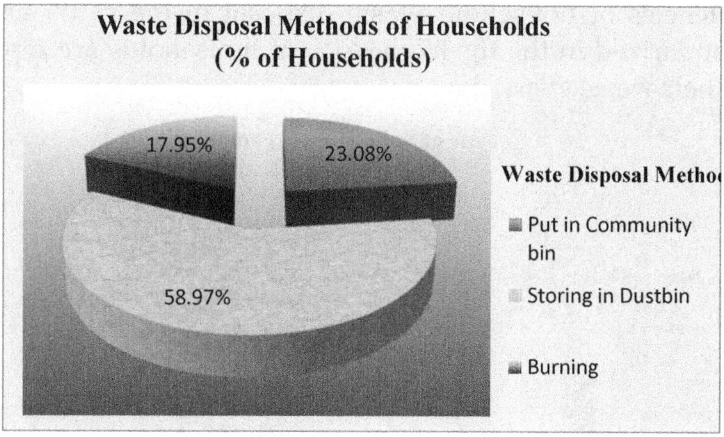

Source: Primary Data, 2020.

Figure 6

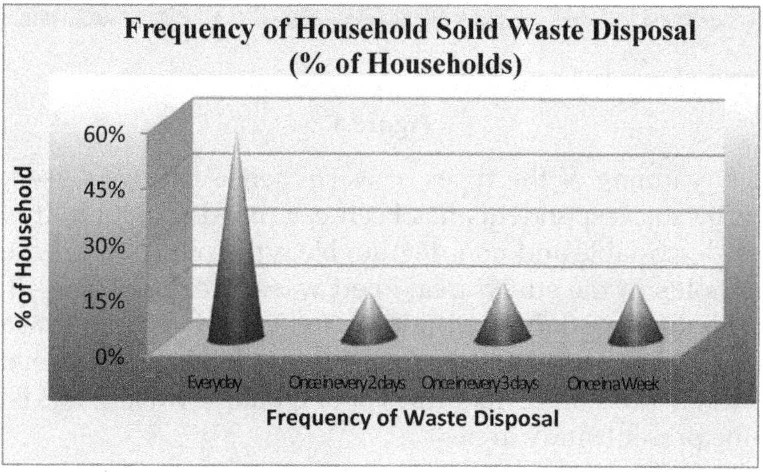

Source: Primary Data, 2020.

Figure 7

Household waste disposal method of the study area has shown in fig 1.6. It is observed that most of the households (58.97%) accumulate their waste in dustbin and more than 23% of households are put their wastes in pubic bin. Therefore, one can come to the conclusion that most of the households have environmental consciousness. The rest only 17.95% of households dispose of their waste simply by burning which indicates they mainly rely upon unscientific method of waste disposal.

Frequencies of household waste disposal in the study area have been summarized in the fig 1.7. 56.41% of households are reported to dispose their waste every day.

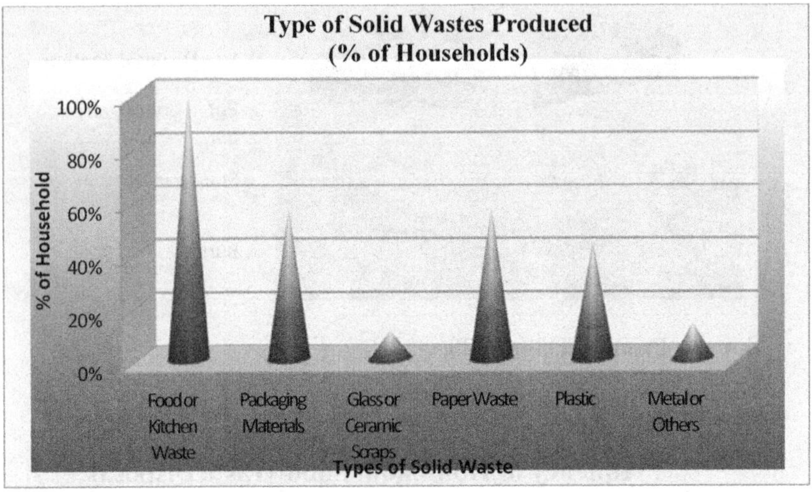

Source: Primary Data, 2020

Figure 8

Quick scanning of the types of waste generated by households as reported by the respondents has been examined in fig 1.8. It is noted that both degradable and non-degradable types of waste are generated by households in the study area. Food waste are generating in all the households of the study area (100%) followed by paper wastes and packaging materials (56.41%). In general household sector in developing countries generates more quantities of degradable wastes; this is true in case of the present study area.

Table 9: Facility of Solid Waste Storage and Disposal

Statements	% of Households	
	Yes	No
Households have storage receptacle for solid waste	84.62%	15.38%
Households segregate different types of waste	53.85%	46.15%
Households have community bins in their locality	51.28%	48.72%
Households practice composting	30.77%	69.23%
Household's awareness about SWM raised by media	46.15%	53.85%

Source: Primary Data, 2020

It is noted from the above table (1.9) that out of 50 respondents more than 84% respondents have storage bins in their household premises. It can be said that respondents in the study area are following scientific method of storing and disposing their household waste. In case of Waste segregation, it is further observed that segregation rate at source is medium accounting for 50%. It is the key principal of solid waste management; it helps to reduce the waste quantity at source, but in Jalpaiguri Municipal area more than 46% of households do not segregate waste. 51.28% of respondents stated that the existence of community bin in the study area. According to them, even though community bins are provided by the municipality in most of the wards, these are largely absent or stolen.

Household Waste Collection Services

Collection of waste is the essential step of solid waste management. It is necessary to ensure that waste stored at source collected regularly and not disposed of on the streets or water bodies. Waste collection services, its frequency, availability, Household's opinion about waste collection services in Jalpaiguri municipal area have been discussed in this section:

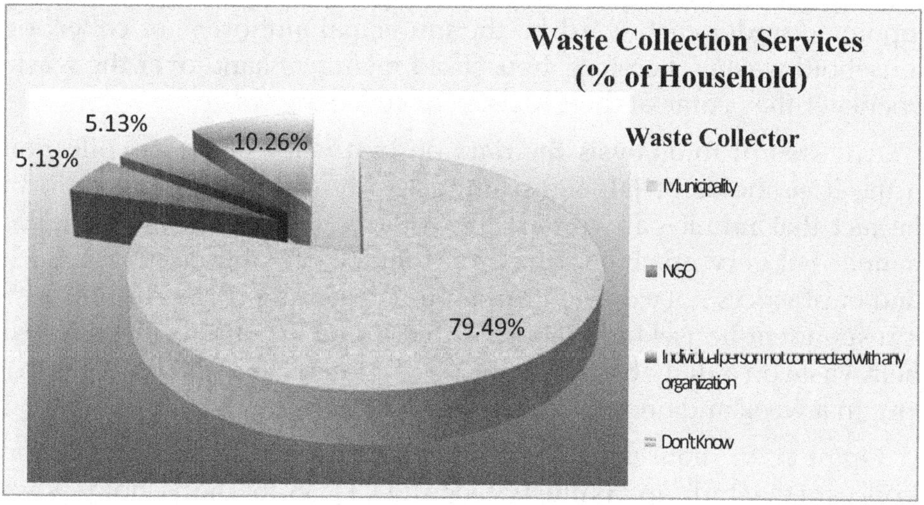

Source: Primary Data, 2020.

Figure 9

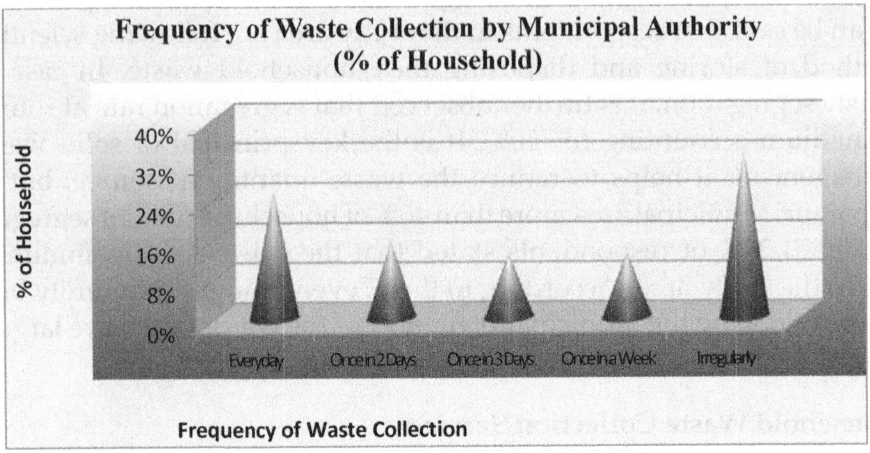

Source: Primary Data, 2020.

Figure 10

In most of towns and cities Municipality has the major responsibility for collecting the waste from household. There is no exception in Jalpaiguri town also. 79.49% of household's waste is collected by municipal authority in the study area (Fig 1.9). Municipal workers are appointed under each ward by the municipal authority for collecting household waste. Therefore, household members hand over the waste whenever they come at their houses for collecting waste.

An attempt to analysis the data on frequency of waste collection in the households of Jalpaiguri municipality (Fig 1.10) brought forth the fact that municipal workers are collects garbage not in a regular manner but very much occasionally from 33.33% of households. Such kind of practices not expected from municipal authorities. Around 25% of respondent households reported that municipal authority collected their waste on a daily basis, followed by once in every 2 days (15.38%), once in a week and once in 3 days (12.82%).

Opinion of household respondents regarding waste collection services of the study area is illustrated in fig 1.11. 38.46% household shows negative responses, when the reason asked for their dissatisfaction, they stated municipal services are not at a satisfactory level, they are irregular and collection workers are impolite, it is really a matter of concern. More than 50% are reported as reasonably satisfied and only 5.13% respondents are fully satisfied with the collection services.

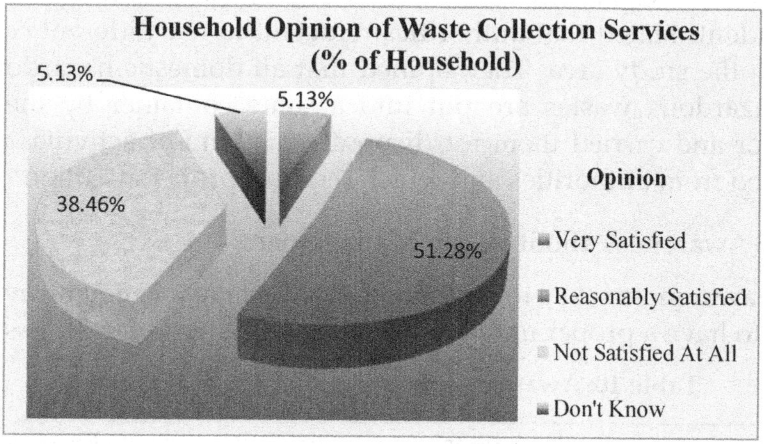

Source: Primary Data, 2020

Figure 11

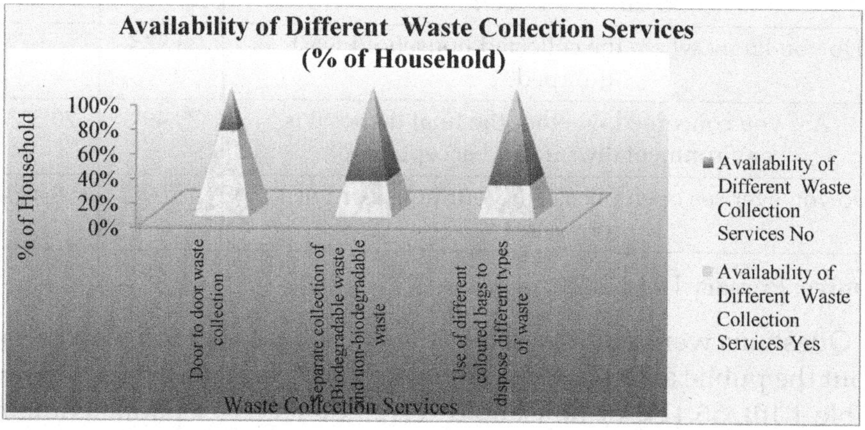

Source: Primary Data, 2020

Figure 12

Respondents are asked to present real picture of waste collection services in the study area so as to assess the efficiency of municipal staffs. The responses are obtained and compiled in fig 1.12 to highlights availability of different waste collection services. 66.67% of respondents reported that door to door waste collection is available in their house. It is conducted in a moderate level in the study area. According to 71.79% of respondents, Municipal authorities do not separate biodegradable waste and non-biodegradable waste at the time of collection, the authority should give more importance in this regard. Colour coded bags help to separate waste more effectively but more than 75% of respondents stated that authorities reluctant to use different coloured bags in the study area. They opined that all domestic hazardous and non-hazardous wastes are put into a same container by the waste collector and carried them for disposal, this kind of activities are not expected from authorities and which requires proper attention.

Public Awareness about Waste Management

Public awareness is key to any successful waste management; everyone needs to have a proper understanding of waste management issues.

Table 10: Awareness about Solid Waste Management

Questions	Responds	
	Yes	No
Do you use a separate container for your kitchen waste and dry waste?	56.41%	43.59%
Do you know where the collected household waste is dumped?	33.33%	66.67%
Are you concerned whether the final disposal is environmentally safe and acceptable?	79.49%	20.51%
Do you ever see open piles near your house or burning of waste?	43.59%	56.41%

Source: Primary Data, 2020

Questions were put forward to the household respondents to know about the public awareness about waste management in the study area (Table 1.10). 56.41% of household respondents use separate container for kitchen and food waste. Majority of them stated that they usually keep two containers one is for plastic, glass, paper and some other kinds of dry waste and the other is for the household kitchen and food waste. Nearly 43.59% of respondents put kitchen and dry waste in a same container, they do not use separate container. The next question was asked whether they know where the collected wastes are being dumped or not, 66.67% respondents answered they are totally unaware about this matter, only 33.33% of respondents gave positive response. An overwhelming percentage (79.49%) of respondents is concerned about the final disposal because they aware about certain waste can harm the environment if not disposed properly. About 43.59% of respondents noticed open piles or burning of waste near their house, it is a matter of great concern because it can create environmental pollution and cause long term health problems of the residents.

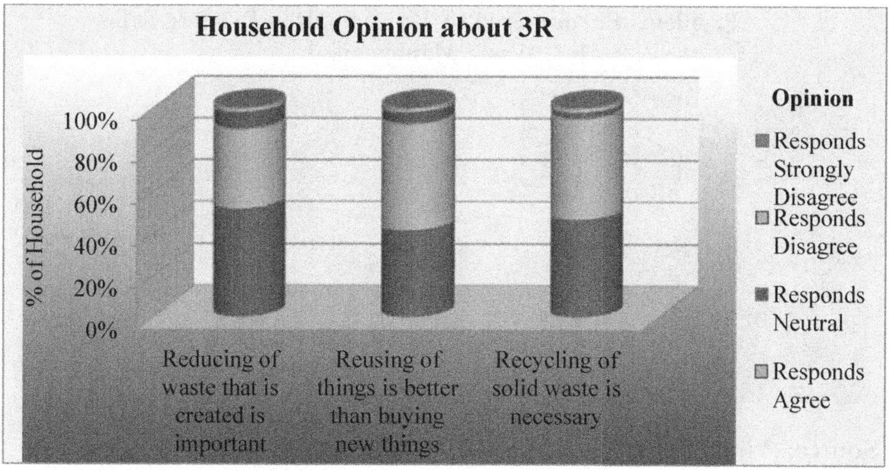

Source: Primary Data, 2020

Figure 13

Respondents opinion about importance of 3R's of solid waste management of the study area are put forth in fig 1.13. These 3R's are-

- **Reduction:** Reduction means the prevention of waste generation itself, so that we can minimize disposal related problem of waste.
- **Reuse:** Reusing helps to recover the item to use again.
- **Recycling:** Recycle changes the recovering the material by changing its form.

It is found in the above table that there are a very few respondents (2.56%) of the study area actually disagreed with this 3R's of waste management. None of the respondents are found to be strongly disagreed. In case of reduction of waste 51.28% are strongly agreed, 38.46% are agreed while 7.69% are remains neutral. Furthermore, 51.28% & 48.72% respondents are reported agreed and 41.03% & 46.15% are strongly agreed in case of reusing and recycling of waste respectively. Therefore, it indicates that the respondents think positively about 3R methods of solid waste management.

Problems and Impact of Untreated Waste

Problems and impact of untreated waste of the study area are discussed and analysed with the help of tables and diagrams:

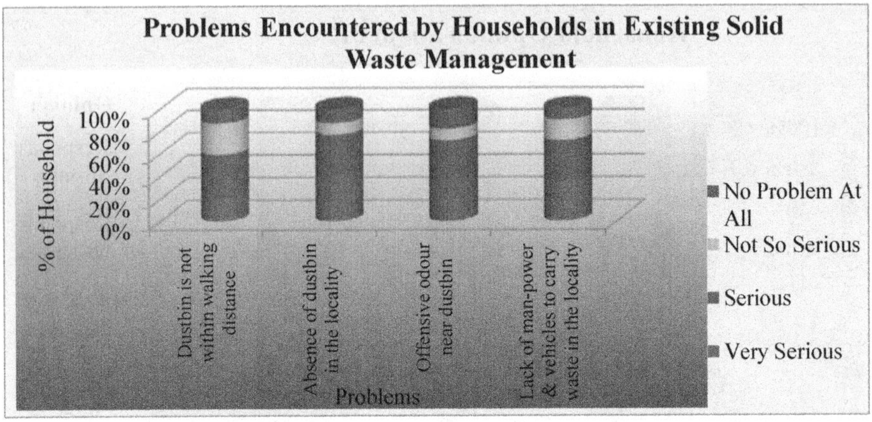

Source: Primary Data, 2020

Figure 14

To evaluate the problems encountered by the households some problems are selected and their responses are recorded (Fig 1.14). It is noticed that location of dustbin and absence of dustbin is a major problem in the area. 38.46% of respondents think that dustbin is not within walking distance is the serious issue in the study area, while 20.51% of them found it very serious and 28.21% said it's not so serious at all. 41.03% of respondents considered absence of dustbins is a very serious problem and nearly 40% of them stated it's a serious issue. Only 12.82% of respondents did not take it as a problem. Therefore, it is worth to mention that dustbins are not equally distributed throughout the Jalpaiguri municipal area. Almost 40% of household respondents stated there is an offensive odour near dustbin this is because municipal authorities have not taken enough care about clearing the dustbins in a regular manner. According to respondents sometimes dustbins are overflowed and still people dump on it. Another major problem is lack of manpower and vehicles to carry out the waste in the study area. 35.90% people found it both very serious and serious problems. When the reason asked, they stated the waste from the household is transported in outdated vehicles specifically in open tractors and trucks. Wastes are often fall off those vehicles at every bump or potholes thus polluting the environment. The municipal authorities fail to adequately provide specially designed covered trucks for waste transport in the study area.

Table 11: Household Suffered from Diseases Due to Untreated Waste

Yes	No
30.77%	69.23%
1. Typhoid fever, 2. Diarrhoea, 3. Malaria, 4. Cholera, 5. Jaundice, 6. Encephalitis, 7. Dengue fever	–

Source: Primary Data, 2020

Household suffered from diseases due to untreated wastes are highlighted in table 1.11. It is observed that most of the respondents (69.23%) do not suffered from any diseases due to untreated waste because they disposed of household waste in a careful manner and their locality is cleaned regularly. 30.77% are reported facing problems and suffered various waterborne diseases due to untreated waste. this is mainly due to poor handling of waste in their locality.

Management of Untreated Waste

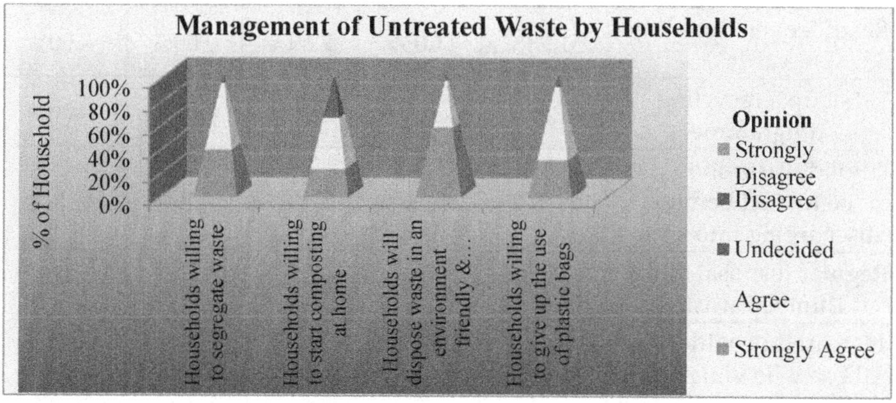

Source: Primary Data, 2020

Figure 15

The most encouraging fact (Fig 1.15) is that 53.85% of respondents are agreed to segregate their waste at home this highlights the awareness ratio of the respondents. 38.46% of respondents are strongly agreed while negligible proportion of respondents showed their ignorance. 41.03% of respondents agree to start composting, 23.08% are strongly agreed, a considerable proportion of respondents do not decide to start composting, they said as they do not segregate their household waste into biodegradable and non-biodegradable items composting is not possible. A considerable number of respondents both strongly agreed and agreed to follow scientific method of waste disposal, they are preparing to dispose of household waste in a manner that is environmentally and socially acceptable. No respondents are found strongly disagreed with this statement. 58.97% of respondents are agreed to give up the use of poly bags, 30.77% of them are strongly agreed, and they are savvy enough to know the adverse effect of using poly bags. 7.69% of respondents remain undecided while 2.56% of them are refused to give up the use of plastics; they opined it is practically impossible not to use plastic bags for their grocery shopping. Therefore, it can be stated that the respondents in the study area in general had good cognition regarding their household waste management.

Table 12: Household's Attitude Towards Following Ways of Solid Waste Reducing Pollution in Study Area

Statements	Strongly Support	Support	Neutral	Oppose	Strongly Oppose
Restrict open dumping yards	51.28%	41.03%	7.69%	0.00%	0.00%
Set up a recycling programme	51.28%	43.59%	5.13%	0.00%	0.00%
Solid waste should be treated before discharging into river	58.97%	35.90%	2.56%	2.56%	0.00%
Regular disposal of the dumped waste	51.28%	43.59%	5.13%	0.00%	0.00%
Increased penalties for those who violating regulations of municipal corporation	46.15%	43.59%	7.69%	2.56%	0.00%
Construction of landfills	38.46%	43.59%	17.95%	0.00%	0.00%

Source: Primary Data, 2020

Open dumping is a common practice in Jalpaiguri municipal area. The largest proportion of respondents (strongly support 51.28% and support 41.03%) are against the open dumping practices in the study area, 7.69% of respondents did not show any interests whereas none of them are strongly oppose with this statement. There is an absence of formal recycling mechanism makes it difficult for everyone to recycle in the study area. Although 51.28% respondents are strongly supporting, 43.59% are supports, 5.13% are neither supports nor oppose for setting up a recycle programme in the study area. Most of the respondents acknowledge the severity of impact of untreated waste so that 58.97% and 35.90% of respondents are strongly supports and supports respectively with the statement that solid waste must be treated first before discharging into the river. Similarly, 51.28% respondents strongly support that regular disposal of waste is necessary, around 44% of respondents are supports while 5.13% of respondents remains neutral. It is interesting to note that more than 40% of the respondents felt that government should increase penalties in any form of public littering and for those who violates the regulations of Municipal Corporation in the study area. In case of construction of landfills around 44% of respondents' shows favourable attitudes as Jalpaiguri municipality have no fix landfill sites as its own for waste disposal.

After analysing the responses of the respondents, it is clearly exhibit that, majority of respondents are prepared to cooperate with the municipal authorities in solid waste management in the study area.

Factor analysis is a statistical technique for identifying which underlying factors are measured by a number of observed variables. Regarding the satisfaction level of the respondent in the study area, several selected parameters have been included and the result has been defined based on the satisfactory parameter level from 1 to 5. Higher values always indicate more positive sentiments ("1" = strongly disagree, "2" = disagree, "3" = neutral, "4" = agree & "5" = strongly agree). All variables have a value "6" which need to set as "no answer" for missing values.

Table 13: Communalities

	Initial	Extraction
Segregation of waste	1.000	0.804
Composting of waste at home	1.000	0.530
Give up use of plastic bags	1.000	0.828
Restrict open dumping of waste	1.000	0.768

	Initial	Extraction
Treatment of waste before discharging into river	1.000	0.630
Solid waste put water resources at risk of contamination	1.000	0.547
Solid waste creastes foul smell on roadside	1.000	0.747
Solid waste heaped at places often attracts insects	1.000	0.614
Open burning of waste creates air pollution	1.000	0.723
Non-biodegradable waste decomposes slowly creates soil pollution	1.000	0.534
Extraction Method: Principal Component Analysis.		

Source: Collected from primary data and computed by the authors

Table 14: Total Variance Explained

Compo-nent	Initial Eigenvalues			Rotation Sums of Squared Loadings		
	Total	% of Variance	Cumulative %	Total	% of Variance	Cumulative %
1	1.801	18.013	18.013	1.534	15.340	15.340
2	1.367	13.671	31.685	1.401	14.010	29.350
3	1.316	13.161	44.846	1.280	12.804	42.154
4	1.201	12.011	56.857	1.271	12.708	54.862
5	1.040	10.398	67.255	1.239	12.393	67.255
6	0.858	8.582	75.836			
7	0.787	7.870	83.707			
8	0.693	6.931	90.638			
9	0.503	5.029	95.666			
10	0.433	4.334	100.000			
Extraction Method: Principal Component Analysis.						

Source: Collected from primary data and computed by the authors

According to the statistical representation like principal component analysis, communality value can be calculated and based on the extracted value it is found that segregation of waste, give up use of plastic bags, restrict open dumping of waste, solid waste creates foul smell on roadside, open burning of waste creates air pollution etc. parameters have been response properly by the respondent of the study area. However, due to solid waste which can make huge contamination in water as well as non-biodegradable waste which creates social pollution are not been included so much because of less extracted value based on the communality calculation. From the table where total variance has explained here 5 components, targeted more than 1 value which explains almost 67% of the cumulative variance explained from the Scree plot. First 5 components make a declining trend line which perform relational diagram between component number and eigenvalue. Therefore 6 to 10 component number it almost performs similarity i.e. why it does not explain the percentage of variance.

Table 15: Rotated Component Matrix[a]

	Component				
	1	2	3	4	5
Segregation of waste	.059	-.105	-.026	.887	-.037
Composting of waste at home	.547	.141	.314	.219	-.254
Give up use of plastic bags	.005	.120	.901	.033	-.002
Restrict open dumping of waste	-.007	.768	.036	-.384	-.172
Treatment of waste before discharging into river	-.149	-.162	-.117	-.285	.698
Solid waste put water resources at risk of contamination	.686	-.188	-.122	.135	.088
Solid waste creastes foul smell on roadside	.201	.213	.118	.230	.771
Solid waste heaped at places often attracts insects	.775	-.054	-.017	-.092	.038
Open burning of waste creates air pollution	-.130	-.521	.560	-.345	.045
Non-biodegradable waste decomposes slowly creates soil pollution	-.283	.620	.107	.083	.226
Extraction Method: Principal Component Analysis. Rotation Method: Varimax with Kaiser Normalization.[a]					
a. Rotation converged in 9 iterations.					

Source: Collected from primary data and computed by the authors.

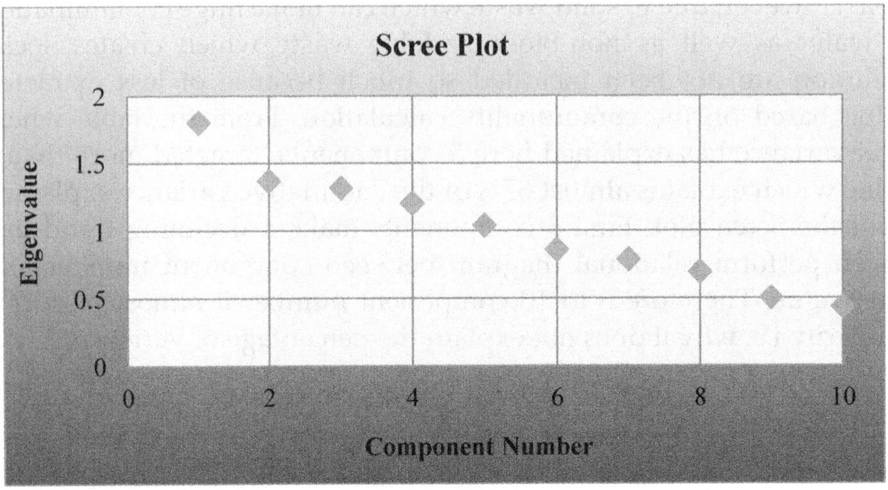

Figure 16

Source: Collected from primary data and represented by the authors

From the given calculation of rotated component matrix (Table 1.13) under principal component 1 segregation of waste, water resource contamination due to solid waste, places attract insects, composting of waste at home are the major common problems which have been derived quite prominently by the calculation. Under principal component 2 restrict open dumping of waste, composting waste at home, non-biodegradable of waste decomposes slowly creates soil pollution are the tree major pertinent factors have been calculated. Under principal component 3 give up use of plastic bags followed by treatment of waste before discharging into river creates foul smell from waste are the major calculated problems which have been given by the respondent. Under principal component 5 segregation of waste are the major common problem derived by the calculation. Lastly under principal component 5 water contamination due to waste disposal, foul smell from solid waste on roadside again another two-problem found in the study area.

Findings and Problems

- The number of the community bins and dustbins are not adequate for the much of population size in the study area. There is no proper allotment of Dustbins; these are not within walk able distance from the households. Most of the community bins spreads odour.

- Door to door collection service is available most of the households but municipal workers are very much irregular at performing their services. Household waste collection frequency varies from daily to about once a week. The workers who are engaged for waste collection are not well trained.
- Collected waste from the households is mostly transported in uncovered or open vehicles. There is absence of modern equipment's and proper waste carrying vehicles in the study area.
- Segregation of waste into bio-degradable and non-biodegradable items at household level is sporadic in the study area; however, some amount of segregation is undertaken by some individual household. Besides, Composting is not a standard practice and it is also confined to limited individuals.
- Methods of disposing of household waste like storing waste in dustbin, put the waste in public bin, burning of waste are very common in the study area. Out of 50 selected households none of the households is found to throw their waste on the street. Therefore, it can be said that the households have great intention in disposing waste and they have environmental consciousness.
- One of the major finding of the study area is that, the households are aware about the solid waste management problems and also very much concerned about lack of proper system of waste collection. People are willing to co-operate with the solid waste management programme if appropriate technology and technical support are provided.
- Most of the households aware about the importance of 3R method of waste management. No recycling programme has been taken up in the study area till now.
- Private voluntary organizations like NGO plays very little role regarding waste management in the study area.

Recommendations

- Focus should be given on developing community level small scale composting process. Large scale composting may be expensive and may not work in a time like Jalpaiguri. Hence, community level composting can be undertaken at local level, it will be easier to manage.

- Creation of three stream waste segregation system such as bio degradable, recyclables, garbage waste will ensure long time sustainability of solid waste management. Waste segregation must be performed at source itself. Segregated waste can be collected daily basis from households.
- The collected household waste should be carried out using covering trucks and trailers. Care should be given not to spill the waste during transportation.
- Emphasis should be placed on reduction, reuse, recycling of waste. These will create less waste and increase material recovery. Involvement of informal sector in recycle activity should be promoted in the study area.
- A sanitary landfill designed specifically for the final disposal should be built. Jalpaiguri municipality should find a proper place in this regard. A cluster-based project has been taken up by Siliguri Jalpaiguri Development Authority for this purpose. Municipal authority should also provide adequate bins to keep the environment free from waste.
- Social organizations and institutions such as schools, colleges, involvement of NGOs should spread awareness and information on the status of waste management - generation, collection, disposal etc. Regular ward level meeting should organise to keep people informed. Research and development should be encouraged.

Conclusion

The study focused on household solid waste generation, disposal, transportation, treatment, attitude and perception of people towards waste management in Jalpaiguri municipal area. Prior to this study, no particular research had been carried out in terms of solid waste management at household level in Jalpaiguri municipal area; this study has partially filled the gap. Through the study of various households, it is identified that most of the respondent households stored their waste either in the dustbins of their household premises or put them in the public bins. Waste collection services are found to be infrequent and inadequate. Household waste transportation services are rudimentary and poses environmental risk, Composting and segregation are not in a regular practice however some household practices by their individual interest. There is no provision for waste segregation, composting and recycling by the Jalpaiguri municipality. Solid waste is collected from households without separation at source, treatment facilities are limited. Household members in general are open to share their responsibilities and they see a role for themselves for waste management.

References

Carter Harold (1995): The Study of Urban Geography, Edward Arnold (co-published by Halsted Press, London, United Kingdom.

Emery Andrew, Davies Anthony, Griffiths Anthony, Williams Keith (2006): Environmental and Economic Modelling: A Case Study of Municipal Solid Waste Management Scenarios in Wales, Resources, Conservation and Recycling 49 (2007) 244-263, ScienceDirect, Elsevier.

Guerrero Lilliana Abarca, Maas Ger, Hogland William (2012): Solid Waste Management Challenges for Cities in Developing Countries, Waste Management, Elsevier.

Miezah Kodwo, Obiri-Danso Kwasi, Kadar Zsofia, Fei-Baffoe Bernard, Y.Mensah Moses (2015): Municipal Solid Waste Characterization and Quantification as a Measure towards Effective Waste Management in Ghana, Waste Management, Elsevier.

Ministry of Environment, Forest and Climate Change (2016): The Gazette of India: New Delhi, India.

Pradhan Upendra Mani (2008): Sustainable Solid Waste Management in a Mountain Ecosystem: Darjeeling, West Bengal, India, Natural Resources Management, Environment Earth and Resources, University of Manitoba, Winnipeg, Manitoba.

Rathi Sarika (2005): Alternative Approaches for Better Municipal Solid Waste Management in Mumbai, India, Waste Management 26 (2006) 1192-1200, ScienceDirect, Elsevier.

Saxena H.M (2010): Environmental Management, Rawat Publications, Jaipur, India.

Talyan Vikash, Dahiya R.P, Sreekrishnan T.R (2007): State of Municipal Solid Waste Management in Delhi, the Capital of India, Waste Management 28 (2008) 1276-1287, ScienceDirect, Elsevier.

Urban Development & Municipal Affairs Department, Government of West Bengal: State Policy and Strategy on Solid Waste Management.

West Bengal Pollution Control Board, Government of West Bengal (2016): Annual Report 2015-2016: Kolkata, India.

List of Contributors

1 **Ananya Chowdhury**

Assistant Teacher, Swarupganj Jr. High School,
Malda, West Bengal, India.

2 **Biswajit Majumdar**

Assistant Teacher, Fatepur C.D. High School,
Kaliyaganj, Uttar Dinajpur. West Bengal, India.

3 **Chandra Kanta Saha**

Assistant Professor and Head, Department of Economics,
Sister Nibedita Government General Degree College for Girls,
Kolkata, West Bengal, India.

4 **Debarati Chowdhury**

Ph.D. Research Scholar,
Department of Geography & Applied Geography,
University of North Bengal, Siliguri, West Bengal, India.

5 **Dr. Abhijit Bhattacharjee**

Assistant Professor, Jalpaiguri Law College,
Jalpaiguri, West Bengal, India.

6 **Dr. Anirban Roy**

Assistant Professor and Head,
Department of Geography & Environmental Science, Government
General Degree College, Paschim Medinipur, West Bengal, India.

7 **Dr. Arindam Basak**

Assistant Professor
Department of Geography & Applied Geography
University of North Bengal, Siliguri, West Bengal, India.

8 **Dr. Dipak Kumar Singh**

Assistant Professor, Department of Education,
Ananda Chandra Training College, Jalpaiguri, West Bengal. India.

9 **Dr. Indrajit Roy Chowdhury**

Assistant Professor,
Department of Geography & Applied Geography,
University of North Bengal, Siliguri, West Bengal, India.

10 **Dr. Krishna Das**

Assistant Professor, Department of Geography,
Pragjyotish College, Guwahati, Assam, India.

11 **Dr. Madhusudan Karmakar**

Assistant Professor, Department of Geography,
Maynaguri College, Jalpaiguri, West Bengal, India.

12 **Dr. Malay Ganguli**

Assistant Professor in Geography (W.B.E.S.),
Sister Nibedita Government General Degree College for Girls,
Hastings House, Kolkata, West Bengal, India.

13 **Dr. Niraj Kumar**

Assistant Teacher, Department of Geography,
S.D. (P.G.) College, Muzaffarnagar, Uttar Pradesh, India.

14 **Dr. Ranjan Sarkar**

Assistant Professor, Department of Geography,
Ananda Chandra Training College, Jalpaiguri, West Bengal, India.

15 **Dr. Ratula Mukherjee**

Assistant Professor, Department of Botany,
Ananda Chandra Training College, Jalpaiguri, West Bengal. India.

16 **Dr. Savita Mishra**

Principal, Vidyasagar College of Education,
Phansidewa, Darjeeling, West Bengal, India.

17 **Dr. Sikha Rani Nath**

Assistant Teacher, Assam Govt. Service,
Guwahati, Assam, India.

18 **Dr. Sudhir Kumar Kisku**

Assistant Professor, Department of Chemistry,
M.P.C Autonomous College, Baripada, Odisha, India.

19 **Gopal Das**

Assistant Teacher in Geography,
Maynaguri Subhas Nagar High School,
Maynaguri, Jalpaiguri, West Bengal, India.

20 **Kanchan Devi**

M.Phil. Research Scholar, Department of Economics,
Rajiv Gandhi University, Arunachal Pradesh, India.

21 **Kantisree Goswami**

Assistant Professor, Department of Zoology,
Charuchandra College, Kolkata, West Bengal, India.

22 **Keerthana. S**

Nutrition and Dietetics, Vellalar College for Women
(Autonomous), Erode, Tamil Nadu, India.

23 **Mayukhmala Mandal**

Assistant Professor, Department of Zoology,
Charuchandra College, Kolkata, West Bengal, India.

24 **Mominul Sinan**

Assistant Professor, Department of Chemistry,
Charuchandra College, Kolkata, West Bengal, India.

25 **Parimita Saikia**

Ph.D. Research Scholar, Department of Geography,
The Assam Royal Global University,
Betkuchi, Guwahati, Assam India.

26 **Partha Das**

Assistant Professor, Department of Geography,
A. B. N. Seal College, Cooch Behar, West Bengal, India.

27 **Soumya Pratik Dutta**

Ph.D. Research Scholar, Department of Law,
Assam University, Silchar, Assam, India.

28 **Subhasmita Parida**

Ph.D. Research Scholar, P.G. Department of Geography,
Utkal University, Vhani Vihar, Odisha, India.

29 **Swati Sharma**

Former Research Scholar, M.Phil., Department of History,
Babasaheb Bhimrao Ambedkar University, Lucknow, Uttar
Pradesh, India.

30 **Tuhin Dey Roy**

Ph.D. Research Scholar,
Department of Geography & Applied Geography,
University of North Bengal, Siliguri, West Bengal, India.

Dr. Ranjan Sarkar's Other Published Books

Socio-Economic Development and Environmental Sustainability: The Indian Perspective

About the Book

The book "The Socio-Economic Development and Environmental Sustainability: The Indian Perspective" presents a multi-dimensional facet of Socio-Economic Development and Environmental Sustainability in Indian Context. The book consists of three parts under which a total number of 25 research articles are there. In the first part, 10 articles related to Man, Environment and Sustainability have been accommodated. Articles on Demographic Trends and Urbanization have been included in the second part while in the third part 10 articles have been incorporated on aspects of Socio-economic Development. In this book emphasis has been given on the various dimensions of socio-economic development and its regional disparity in a planned way. Man depends on environment for his sustainability. The relationship between environment and man is indeed very deep and has been recognized right from ancient time. Over a period of time man's changing social environment has been modified by population growth, urbanization and industrialization. This in turn has affected the natural environment to a great extent. Man's increasing demands and technological developments has made him irresponsible in his attitude towards nature and its resources. Large scale deforestation, soil erosion, pollution and the extinction of species are only some of the outcomes of his exploitative abuse of the natural environment. Pollution problems have created in serious health problems, as well as negative impact on the environment. The present volume is associated with such problems and accounted for the suitable measures to unlock a new direction towards balanced Economic growth and Environmental Sustainability.

Geography in the 21st Century: Emerging Issues and the Way Forward

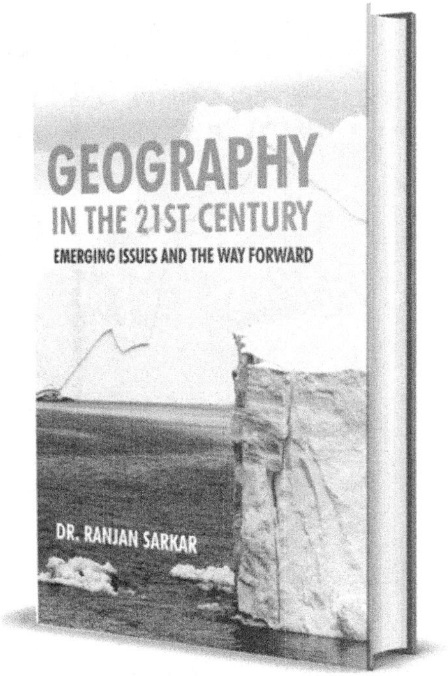

About the Book

Geography is considered as mother of all sciences and this is one of the oldest disciplines in the World. As a scientific discipline, Geography has always embraced new technologies, research practices, instructional methods, skills and content. Learning Geography is creative and can inspire interest about the world that involves a curiosity about nature. The book entitled 'Geography in the 21st Century: Emerging Issues and the Way Forward' represents the various emerging issues from varied branches of Geography from traditional to modern perspectives in a planned way. It is divided into three sections: Population, Migration and Changing Landscape; Dynamics of Socio-Economic Development; and Prospective Trends in Geography. In this book there are 25 chapters which have been contributed by various established academicians and researchers across India. They lucidly highlighted and expressed their research-oriented ideas, views and outcomes regarding various issues on Population, Migration and Changing Landscapes; about dynamics of Socio-economic Development and prospective trends in Geography along with the modern tools and techniques for the purpose of betterment of the society. This book will surely be beneficial to the policy makers, academicians and researchers of Social Sciences in general and Geography in particular.

Education in the 21st Century: Emerging Issues and the Way Forward

About the Book

Education is one of the most powerful tools which can be used to change the world. It helps in the development of critical thinking which is necessary for the development of scientific temperament. Education in the 21st century is going through a paradigm shift. This century has witnessed the journey from the traditional face to face classroom to the modern virtual classroom. The goal of modern education is to focus on ensuring that children would be problem solvers, decision makers and enablers. Education in present digital world is concerned with giving learners the skills they need to succeed in this new world and helping them grow the confidence to practice those skills. Students need to leave school with life skills that help them navigate challenges, even if they don't know the solutions to them. The book entitled 'Education in the 21st Century: Emerging Issues and the Way Forward' represents the diverse issues, trends and challenges relevant to the present education system. It is divided into three sections: Modern Approaches to Education; Educational Awareness and Development; Literacy in the Present context and Its Disparity. In this book total 20 chapters have been presented to depict the major concerns related to education.

Dr. Ranjan Sarkar, M.A. (Geography-Gold Medalist), B.Ed., M.Ed., NET, Ph.D. (Geography), is an eminent academician. He has completed his M.A. Degree in Geography & Applied Geography from the University of North Bengal, Siliguri, West Bengal, India, in the Year 2003 and secured 1st Class 1st Position. He has Successfully qualified University Grants Commission (UGC)-NET in June,2004. He also obtained B.Ed. and M.Ed. degrees. He has been awarded Ph.D. Degree in Geography & Applied Geography from the University of North Bengal. He is the recipient of North Bengal University Gold Medal in Geography-2003, Bidhu Ranjan Adhikary Memorial Trust Award-2003, Dr. S.B.Chatterjee Memorial Trust Award-2002, National Talent Search Award-1995. Since 2006, he has been working in the post of Assistant Professor in Geography (Senior Scale) of Ananda Chandra Training College, (Through The West Bengal College Service Commission), Jalpaiguri, West Bengal, India. Before that he had served as an Assistant Teacher in Geography for H.S. Section in Kushkari K.B.B.S. High School (H.S.), Kushkari, Dakshin Dinajpur, West Bengal (Through The West Bengal Central School Service Commission). He has been acted as Nominated Member, Governing Body of various Colleges, Selection Committees, NAAC Coordinator, Academic Counsellor-Netaji Subhas Open University (NSOU) & Indira Gandhi National Open University (IGNOU), DIET Subject expert and so on. He has participated more than 110 Regional, National and International Seminars, Workshops, Conferences, Webinars and UGC-HRDC courses actively and Chaired numerous Technical Sessions/Occasions. He has also contributed more than 20 research papers in UGC Approved/Enlisted/CARE National, and International Journals of Repute and edited volumes with ISBN.He has Edited and Published Eight Books with Separate ISBN as Hard Back and E-Book on 'Socio-Economic Development and Environmental Sustainability: The Indian Perspective', 'Geography in the 21st Century: Emerging Issues and The Way Forward', 'Environmental Sustainability in the 21st Century: Emerging Issues and The Way Forward' and 'Education in the 21st Century: Emerging Issues and The Way Forward'.His research areas of interest are Socio-Economic Development, Population Studies, Environmental Sustainability, Technology in Education and Teaching, Modern Trends in Education, Digital Learning and Teaching, Pedagogy of Teaching, ICT. etc. He is an Effective Teacher, Academic Writer, Motivational Speaker, Academic Counselor and Expert Consultant in Educational Supervision and Administration.

NAMYA PRESS.com

₹ 1495/-

ISBN: 978-81-947715-1-7

9 788194 771517

Made in the USA
Monee, IL
04 November 2026

67384030R00203